Cambridge Texts in Applied Mathematics

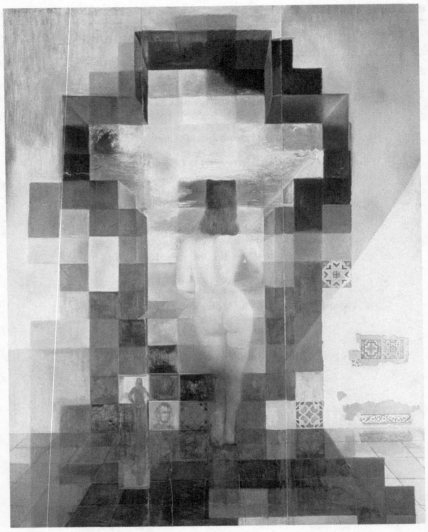

Gala Contemplating the Mediterranean Sea (detail). © Salvador Dali, Gala-Salvador Dali Foundation, DACS, London, 2003. Image supplied by Bridgeman Art Gallery. One of the most important concepts presented in this book is that of intermediate asymptotics. It is illustrated in chapter 2, Figure 2.3, by a tiled version of the photograph of Abraham Lincoln on a $5 bill (Harmon 1973). The paper by Harmon, and, in particular, this tiled picture inspired Salvador Dali to create in 1976 the painting presented here, where some tiles are themselves pictures: of his wife Gala entering the sea, Harmon's original tiled picture of Lincoln, and others. This painting is in fact an excellent example of multiscale intermediate asymptotics.

SCALING

GRIGORY ISAAKOVICH BARENBLATT, For.Mem.RS

*Professor-in-Residence at the University of California at Berkeley,
and Lawrence Berkeley National Laboratory,
Emeritus G.I. Taylor Professor of Fluid Mechanics at the
University of Cambridge,
Adviser, Institute of Oceanology, Russian Academy of Sciences,
Honorary Fellow, Gonville and Caius College, Cambridge*

CAMBRIDGE
UNIVERSITY PRESS

CAMBRIDGE UNIVERSITY PRESS
Cambridge, New York, Melbourne, Madrid, Cape Town, Singapore,
São Paulo, Delhi, Dubai, Tokyo, Mexico City

Cambridge University Press
The Edinburgh Building, Cambridge CB2 8RU, UK

Published in the United States of America by Cambridge University Press, New York

www.cambridge.org
Information on this title: www.cambridge.org/9780521533942

First published 2003
Reprinted with corrections 2010

A catalogue record for this publication is available from the British Library

ISBN 978-0-521-82657-8 Hardback
ISBN 978-0-521-53394-2 Paperback

In grateful memory of my dear friends
Yakov Borisovich Zeldovich and Alexandr Solomonovich Kompaneets

Contents

Foreword

For the past seven years students and faculty at the University of California at Berkeley have had the privilege of attending lectures by Professor G.I. Barenblatt on mechanics and related topics; the present book, which grew out of some of these lectures, extends the privilege to a wider audience. Professor Barenblatt explains here how to construct and understand self-similar solutions of various physical problems, i.e. solutions whose structure recurs over differing length or time scales and different parameter ranges. Such solutions are often the key to understanding complex phenomena; there is no universal recipe for finding them, but the tools that can be useful, including dimensional analysis and nonlinear eigenvalue problems, are explained here with admirable conciseness and clarity, together with some of the multifarious uses of self-similarity in intermediate asymptotics and their connection with wave propagation and the renormalization group. Whenever possible, Professor Barenblatt shuns dry and distant abstraction in favor of the telling example from his incomparable stock of such examples; with the appearance of this book, there is no longer any excuse for any scientist not to master these simple, elegant, crucial and sometimes surprising ideas.

This book is also very timely. Dimensional analysis and simple similarity arguments (what is called here complete similarity) are quite familiar to most scientists, with the possible exception of many mathematicians, yet the deeper, more beautiful and exceptionally useful idea of incomplete similarity, with its extraordinary ramifications, is not yet part of everyone's scientific culture. Maybe part of the reason is the absence of a book that is both sound and accessible. After all, the original papers by Barenblatt and Zeldovich and by others were addressed to the expert; the previous books by Professor Barenblatt are rich in theory and examples and therefore not always easy to read; the very interesting book by Goldenfeld on the renormalization group, where the connection with incomplete similarity is carefully explained, assumes a wider knowledge of modern theoretical physics than can be expected from experts in other fields. If this were the only reason for ignorance then the current book would solve the problem: it is accessible and direct and can be read with profit even by undergraduates.

I suspect however that the difficulty in assimilating the notion of incomplete similarity had deeper sources as well: here is a simple mathematical procedure which makes it possible to contemplate, and indeed often rationally analyze, the disquieting possibility that small parameters may have persistent large-scale effects, not confined to the margins of a physical domain as in most textbook examples of singular perturbations, and not safely relegated to the exotic realm of phase transitions and critical phenomena, but observable in simple physical situations. It is natural to resist ideas which fly in the face of comfortable habits of thought, but this would be a big mistake: the possibility is real, and its understanding requires full attention because it is important. Indeed, the major pedagogical outcome of the example involving turbulent boundary layers in Chapter 8 is to show how admitting the possibility of incomplete similarity can lead to conclusions that are innovative, striking and subversive of long-accepted beliefs.

The importance of this range of ideas is now growing fast. We are at the beginning of the age of multiscale science and multiscale computation, with a growing need to understand not only phenomena on each of many scales but also the interaction between phenomena at very different scales; such interactions abound in fields such as materials science and biology and by definition occur when the impact of the parameters that describe small scales propagates across the full range of scales in a problem. Incomplete similarity is the basic paradigm of how such an impact propagates and is a major tool in the analysis and understanding of new classes of problems and in the emerging art of solving them on a computer.

What we have before us is a clear, masterful and uniquely timely book by one of the great applied mathematicians, who brings us his own great knowledge and experience of a key topic and, furthermore, some of the accumulated experience of the great Soviet school of applied mathematics in which he grew up and of which he is the most distinguished living embodiment.

Alexandre Chorin
University Professor
University of California

Preface

Applied mathematics is the art of constructing mathematical models of phenomena in nature, engineering and society. In constructing models it is impossible to take into account all the factors which influence the phenomenon; therefore some of the factors should be neglected, and only those factors which are of crucial importance should be left. So we say that every model is based on a certain *idealization* of the phenomenon. In constructing the idealizations the phenomena under study should be considered at 'intermediate' times and distances (think of the visual arts, and specifically the impressionists!). These distances and times should be sufficiently large for details and features which are of secondary importance to the phenomenon to disappear. At the same time they should be sufficiently small to reveal features of the phenomena which are of basic value. We say therefore that every mathematical model is based on '*intermediate asymptotics*'.[1]

The construction of an appropriate idealization is the most difficult stage in mathematical modelling. It is always performed in steps. Trial and error, and comparison with experiments, physical or computational, play a basic role. The reader can find a truly remarkable and very exact description of this process in a seemingly unexpected place: Maurice Maeterlinck's *Blue Bird* – it was not by accident that this play won Maeterlinck a Nobel Prize.

Mathematics is considered to be the language of science, because its role in constructing mathematical models is similar to that of language in human communications. All people use language. However, among users of language

---

[1] Later we will give a precise definition of this concept. For the time being we note that the writing of *War and Peace* by Leo Tolstoy is a remarkable example of such an intermediate vision. The novel was extemely successful from the very beginning because lesser details of the Napoleonic war had decayed in people's memories whereas gigantic historic events and their influence on human destinies appeared unshadowed both by small details in the past and current events in the life of society. I am afraid that the literature has missed such an opportunity for the 'Thirty Years War', 1914–1945.

a particularly important group can be distinguished. These are the *authors*: poets, novelists, playwrights, essayists etc., who create fictional images and paradigms – idealized models of people and social phenomena around them. The greatest of these paradigms continue to live for centuries and even millenia. They transform human culture and, sometimes, language itself.

To a certain extent a similar role is played by applied mathematicians. Using the language of mathematics, developing and transforming it when necessary,[2] applied mathematicians create their paradigms – models of phenomena. These idealized models should be sufficiently complete images of phenomena, and at the same time they should enable further study – analytic and experimental, computational and physical. The right to existence of these models is determined by one thing only: they have to work, i.e. they must predict the behavior of the systems under study in interesting but as yet unexplored ranges of external conditions. When this goal is achieved, it leads to practical applications.

Of special importance is the following fact: the construction of models, like any genuine art, cannot be taught by reading books and/or journal articles (I assume that there could be exceptions, but they are not known to me). The reason is that in articles and especially in books the 'scaffolding' is removed, and the presentation of results is shown not in the way that they were actually obtained but in a different, perhaps more elegant way. Therefore it is very difficult, if not impossible, to understand the real 'strings' of the work: how the author really came to certain results and how to learn to obtain results on your own.

Therefore, just like in every art an appropriate way to become an applied mathematician is to become part of a good school, i.e. to work for some time in a team under the guidance of a genuine master. Knowledge, experience and inspiration come from constant discussions on scientific and non-scientific matters, not only with the master gender but also with colleagues and other members of the team. Conversations about music, literature, history, visual art etc. are as important in the process of education as are scientific debates (political discussions of all scales are counter-productive). To be part of a school means to live in a unique environment where an intensive flow of ideas is customary. The present author was fortunate to belong to the school of A.N. Kolmogorov, and to work closely for a long time with Ya.B. Zeldovich, after being taught and strongly influenced by his first teacher, the eminent analyst B.M. Levitan. I want to mention here also a remarkable physicist and teacher of

---

[2] I emphasize *when necessary*. To reproach applied mathematicians for using well-known tool of analysis to construct models is ridiculous: it is the same as reproaching Rafael for not inventing new brushes and paints. Leonardo did, and without any improvement to his paintings this made great trouble for future restorers.

physics, A.S. Kompaneets. I owe him so much for my understanding of physics. My gratitude to these outstanding people is immense.

As with every art, constructing intermediate asymptotics and models has many practical devices and tricks. They should be assimilated. Moreover, they should enter the conscience and subconscience of a researcher who has decided to become an applied mathematician. One of them is the ability to extract from available evidence *scaling laws* or *power laws*. One may ask, why is it that scaling laws are of such distinguished importance? The answer is that *scaling laws never appear by accident*. They always manifest a property of a phenomenon of basic importance, 'self-similar' intermediate asymptotic behavior: the phenomenon, so to speak, repeats itself on changing scales. An explanation could be useful here. The statement that a certain phenomenon is steady, i.e. does not depend on time, is obviously rich in content. The self-similarity of a phenomenon that is developing in time means that with an appropriate choice of time-dependent scales, the phenomenon being measured in these scales becomes "frozen," i.e. steady. This behavior should be discovered, if it exists, and its absence should also be recognized. The discovery of scaling laws very often allows an increase, sometimes even a drastic change, in the understanding of not only a single phenomenon but a wide branch of science. Of course, any parameter, not only the time, can play the same role. The self-similarity means that in such cases the scales could be chosen, depending on this parameter, so that in these scales the phenomenon ceases to depend on the parameter. The history of science of the last two centuries knows many such examples.

It is this subject that the present book is about. In writing this book I have followed the rule which I learned from my great mentor A.N. Kolmogorov: never start teaching or research in a new field of applied mathematics from general concepts, statements, theories and theorems. Consider some instructive examples and the general theory will come and be cast naturally.

In Berkeley I delivered, many times, a course of 30 double lectures which closely follows the present book. I expect that this book can be considered as a textbook for graduate and advanced undergraduate courses. Some parts can also be used in courses such as the strength of materials, the theory of elasticity, fracture, electrodynamics, heat and mass transfer, fluid mechanics, the flow of non-Newtonian fluids and many others. However, if the course is to be taken as a whole then my recommendation is to consider the first six-and-a-half chapters (up to the biological example) as mandatory. It will take approximately three-quarters of a semester. The remaining time can be used for the detailed presentation of a particular topic appropriate for the audience. I myself have selected turbulence. To many people the subject presented in the chapter on turbulence, based on our joint work with A.J. Chorin and V.M. Prostokishin, may seem rather controversial, although not to me. This example gives a unique possibility of presenting together general pricincples and the use of freshly obtained large experimental databases.

I have previously written several books about the subject presented here. (I remember with deep gratitude the publisher from 'Gidrometeoizdat', Mrs O.V. Vlasova, Mrs T.G. Nedoshivina, and Mrs L.L. Belen'kaya. They published my first book in Russian in spite of the serious risk of losing their jobs.) Naturally, some material from my earlier books will find its place in the present book too, particularly material regarding dimensional analysis and physical similarity, in only slightly modified form. However, the central part of this book is entirely new: in particular I have replaced some complicated and difficult basic examples with simpler ones.

I want to express my thanks to Cambridge University Press (Dr D. Tranah and Dr A. Harvey). In fact, the very idea that I should write such an 'intermediate' book matching my inaugural lecture (Barenblatt 1994) and the large book (Barenblatt 1996) belongs with these gentlemen.

I want to express my gratitude to Professor V.M. Prostokishin, who attended all my lectures and gave me important advice both about the lectures and the present book. I am grateful to Professor L.C. Evans and Professor M. Brenner for reading the manuscript and for valuable comments. I want to thank Professors S. Kamin, R. Dal Passo, M. Bertsch, N. Goldenfeld, D.D. Joseph, L.A. Peletier, G.I. Sivashinsky and J.L. Vazquez for the stimulating and friendly exchange of thoughts concerning the subjects presented in this book over many years. I thank Mrs Deborah Craig for processing the manuscript.

To my friend Alexandre Chorin I want to express special thanks for our remarkable time in Berkeley. I have learned from him a lot, in particular his basic paradigm of computational science: this is a different, independent and very productive way of mathematical modelling. I hope to be able to use this knowledge in my future work.

Introduction

The term *scaling* is used in multiple branches of human activity: from forestry and dentistry to theoretical physics. Each time it has a different meaning, not always well defined. In the present book *scaling* describes a seemingly very simple situation: the existence of a power-law relationship between certain variables y and $x_1, \ldots, x_k$,

$$y = A x_1^{\alpha_1} \cdots x_k^{\alpha_k} \tag{0.1}$$

where $A, \alpha_1, \ldots, \alpha_k$ are constants. Such relations often appear in the mathematical modelling of various phenomena, not only in physics but also in biology, economics, and engineering. However, scaling laws are not merely some particularly simple cases of more general relations. They are of special and exceptional importance; scaling never appears by accident. Scaling laws always reveal an important property of the phenomenon under consideration: its *self-similarity*. The word 'self-similar' means that a phenomenon reproduces itself on different time and/or space scales – I will explain this later in detail.

I begin with one of the most illuminating examples of the discovery of scaling laws and self-similar phenomena: G.I. Taylor's analysis of the basic intermediate stage of a nuclear explosion. At this stage a very intense shock wave propagates in the atmosphere and the gas motion inside the shock wave can be considered as adiabatic.

This work started in one of the worst and most alarming days of the Battle of Britain, in the early autumn of 1940. Cambridge professor Geoffrey Ingram Taylor was invited to a business lunch at the Athenaeum by Professor George Thomson, chairman of the recently appointed MAUD committee (the name 'MAUD' originally appeared by chance, but later it was interpreted as the acronym for 'military application of uranium detonation'). G.I. Taylor was told that it might be possible to produce a bomb in which a very large amount of energy would be released by nuclear fission – the name 'atomic bomb' had not yet been used. The question was: what mechanical effect might be expected if such an explosion were to occur? The answer would be of crucial importance for the further development of events. Shortly before this conversation the confidential

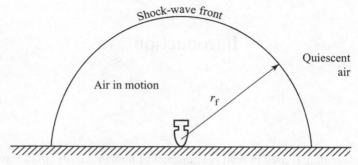

Figure 0.1. A very intense shock wave propagating in quiescent air.

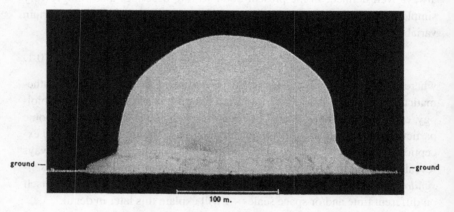

Figure 0.2. Photograph of the fireball of the atomic explosion in New Mexico at
$t = 15$ ms, confirming in general the spherical symmetry of the gas motion (Taylor
1950b, 1963).

report of G.B. Kistyakovsky, the well-known American expert in explosives,
had been received. Kistyakovsky claimed that even if the bomb were successfully constructed and exploded, its mechanical effect would be much less than
expected because the main part of the released energy would be lost to radiation.
As R.W. Clark wrote in his instructive book (Clark 1961), in the whole of Britain
there was only one man able to solve this problem – Professor G.I. Taylor.

To answer this question, G.I. Taylor had to understand and calculate the
motion of the ambient gas after such an explosion. It was clear to him that, after
a very short initial period (related as we now know to thermal-wave propagation
in quiescent air), a very intense shock wave would appear (Figure 0.1). The
motion was assumed to be spherically symmetric, that is, identical for all radii
going out from the explosion centre. (This simplifying assumption later received
excellent confirmation in the first atomic test; see Figure 0.2.) For constructing

a complete mathematical model the following partial differential equations of motion inside the shock wave had to be considered:

1. the equation for the conservation of mass;
2. the equation for the conservation of momentum;
3. the equation for the conservation of energy.

It was intuitively clear to G.I. Taylor that at this early stage in the explosion viscous effects could be neglected and the gas motion could be considered as adiabatic. The above equations of motion had to be supplemented by the following boundary conditions at the shock-wave front:

1. the condition for the conservation of mass;
2. the condition for the conservation of momentum;
3. the condition for the conservation of energy.

Also, the initial conditions, at the beginning of the very intense shock-wave-propagation stage of a nuclear explosion, had to be prescribed.

In fact, this primary mathematical model is so complicated that even now nobody is able to treat it analytically. Adequate computing facilities at that time were non-existent. Moreover, the problem formulation outlined above is incomplete, because nobody knew then or knows now how the air density, air pressure and air velocity are distributed inside the initial shock wave at the time when the shock wave just outstrips the thermal wave and the adiabatic gas motion begins.

G.I. Taylor, however, was astute. His ability to deal with seemingly unsolvable problems, by apparently minor adjustment converting them to problems admitting simple and effective mathematics, was remarkable. And here also he took several steps, of crucial importance, which allowed him to obtain the solution that was needed in a simple and effective form. In addition his formulation allowed him to overcome the lack of detailed knowledge of the initial distribution of the gas density, pressure and velocity. G.I. Taylor's steps were as follows:

1. He replaced the problem by an 'ideal' one. As he wrote (see Taylor 1941, 1950a, 1963), this ideal problem is the following: '*A finite amount of energy is suddenly released in an infinitely concentrated form.*' This means that r_0, the initial radius of the shock wave (the radius at which the shock wave outstrips the thermal wave), is taken as equal to zero, that is, the explosion is considered as instantaneous and coming from a point source of energy. It is clear that neglecting the initial radius of the shock wave r_0 is allowable (if at all!) only when the motion is considered at a stage when the shock front radius r_f is much larger than r_0. If the initial shock-wave

radius is taken as equal zero then the initial distributions of the air density, pressure and velocity inside the initial shock wave disappear from the problem statement: a great simplification.

2. At the same time, he restricted himself to consideration of the motion at the stage when the maximum pressure of the moving gas, reached at the shock-wave front, is large, much larger than the pressure p_0 in the ambient air; this allowed him to neglect the terms involving the initial pressure p_0 in the boundary conditions at the shock-wave front and in the initial conditions. Note that, namely, this stage determines the mechanical effect of the explosion.

The first question G.I. Taylor addressed was: what are the quantities on which the shock-wave radius r_f depends? In the original 'non-ideal' problem they are obviously:

1. E, the total explosion energy, concentrated in the sphere of radius r_0 where the shock wave outstrips the thermal wave (according to the second assumption above the initial internal energy of the ambient quiescent air is negligible);
2. ρ_0, the initial density of the ambient air;
3. t, the time reckoned from the moment of explosion;
4. r_0, the initial radius of the shock wave;
5. p_0, the pressure of the ambient quiescent air;
6. γ, the adiabatic index.

The units for measuring these quantities in the c.g.s. system of units are

$$[E] = \frac{\text{g cm}^2}{\text{s}^2}, [\rho_0] = \frac{\text{g}}{\text{cm}^3}, [t] = \text{s}, [r_0] = \text{cm}, [p_0] = \frac{\text{g}}{\text{cm s}^2}; \qquad (0.2)$$

γ is a dimensionless number. We shall see later how important it was that G.I. Taylor neglected the last two quantities r_0 and p_0, thus replacing the problem by an ideal one.

The reader may ask a natural question: in the real explosion r_0 and p_0 are certain positive numbers which definitely influence the whole gas motion from the very beginning to the end. How can their values be taken to be equal to zero?

In fact (and this comment will be important in our future analysis), the real content of Taylor's assumption was that *at the intermediate stage under consideration, where the mechanical effect occurs,* the motion remains the same if we replace r_0 by λr_0, and p_0 by μp_0. Here λ and μ are arbitrary positive numbers 'of order unity'. This will be explained in detail in Chapter 5, but those who are familiar with the idea of a transformation group even vaguely, will recognize that in fact this was an assumption of group invariance at the all-important intermediate stage.

Taylor's next step can be represented in the following way. He introduced the quantity

$$R = \left(\frac{Et^2}{\rho_0} \right)^{1/5}, \tag{0.3}$$

which is measured according to (0.2) in units of length. Then, if we replace centimeters, cm, by another unit of length, m, mm, μm, km, ..., or in general by cm divided by an arbitrary positive number L, the value of R will be magnified by L, as will also the value of r_f, whereas the quantity

$$I = \frac{r_f}{R} \tag{0.4}$$

obviously will remain unchanged.

The quantity I depends in principle on the same quantities as r_f, and this dependence can be represented, neglecting r_0 and p_0, as

$$I = \frac{r_f}{R} = F(R, \rho_0, t, \gamma) \tag{0.5}$$

where F is a certain function which is not known. The arguments r_0 and p_0 were neglected by Taylor: this was, as we will see, a step of crucial importance. The argument γ is an numerical constant.

The first three arguments of F have independent dimensions. This means, in particular, that time t is measured in time units, i.e., seconds or otherwise s/T where T is an arbitrary positive number. Units of time are absent in the dimensions of the first two arguments; therefore, by varying the number T we can vary the numerical value of the argument t while leaving the values of I and those two other arguments of I invariant (all three others, in fact, since γ is a fixed number). But this means exactly that I cannot depend on t. Similarly with ρ_0: if we vary the unit of mass then the value of ρ_0 will vary arbitrarily, leaving I and the first argument R invariant. That means that I likewise does not depend on ρ_0. Furthermore, I does not depend on the argument R: by varying the unit of length we vary R, but the value of I remains invariant. Thus, the function F is simply a constant depending on the value of γ, and so Taylor's famous scaling law for the radius of the shock wave was obtained:

$$r_f = C(\gamma) \left(\frac{Et^2}{\rho_0} \right)^{1/5}, \tag{0.6}$$

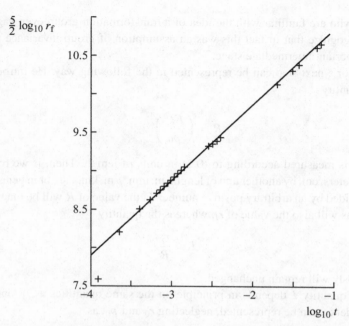

Figure 0.3. Logarithmic plot of the fireball radius, showing that $r_f^{5/2}$ is proportional to the time t (Taylor 1950b, 1963).

or, in the logarithmic form that he used,

$$\frac{5}{2} \log_{10} r_f = \frac{5}{2} \log_{10} C + \frac{1}{2} \log_{10} \left(\frac{E}{\rho_0} \right) + \log_{10} t. \tag{0.7}$$

Later, Taylor's processing of the photographs taken by J.E. Mack of the first atomic explosion in New Mexico in July 1945 (Taylor 1950b, 1963) confirmed this scaling law (Figures 0.2 and 0.3) – a well-deserved triumph of Taylor's intuition. We can see how important it was to neglect the arguments r_0 and p_0, the initial radius of the shock wave and the initial pressure. If not, additional variable arguments would have appeared in the function F and we would have returned to the hopeless mathematical model that we faced at the outset. But the outcome for the simplified situation was different. Taylor was able to obtain in the same way scaling laws for the pressure, velocity and density immediately behind the shock-wave front:

$$p_f = C_p(\gamma) \left(\frac{E^2 \rho_0^3}{t^6} \right)^{1/5}, \qquad \rho_f = C_\rho(\gamma) \rho_0, \qquad u_f = C_u(\gamma) \left(\frac{E}{t^3 \rho_0} \right)^{1/5}. \tag{0.8}$$

Inside the shock wave an additional argument, the distance r from the center of the explosion, appears, so that the relationships for the pressure, density and velocity inside the shock wave can be represented in the form

$$p = p_f P\left(\frac{r}{r_f}, \gamma\right), \qquad \rho = \rho_f R\left(\frac{r}{r_f}, \gamma\right), \qquad u = u_f V\left(\frac{r}{r_f}, \gamma\right). \qquad (0.9)$$

The structure of the relationships (0.9) obtained by Taylor is instructive. It demonstrates that the phenomenon has the important property of *self-similarity*. This means that the spatial distribution of pressure (and other quantities) varies with time while remaining always geometrically similar to itself (Figure 0.4(a)): the distribution at any time can be obtained from that at a different time by a simple similarity transformation. Therefore in 'reduced' coordinates using p_f, ρ_f, u_f and r_f as corresponding scales,

$$\frac{p}{p_f}, \qquad \frac{\rho}{\rho_f}, \qquad \frac{u}{u_f}, \qquad \text{and} \quad \frac{r}{r_f},$$

the spatial distributions of pressure, density and velocity remain invariant in time (Figure 0.4(b)). The property of self-similarity greatly simplifies the investigation: instead of the two independent variables r and t in the system of differential equations, boundary conditions and initial conditions mentioned above, Taylor obtained one single variable argument, r/r_f, in his solution and so was able to reduce the original problem, which required the solution of partial differential equations to the solution of a set of ordinary differential equations. The method of solution was sufficiently simple that he himself was able to make all the necessary numerical computations using a primitive calculator. In particular, he showed that the constant C in the scaling law (0.6) is close to unity: for $\gamma = 1.4$, $C = 1.033$.

G.I. Taylor submitted his paper on Friday 27 June 1941. The great American mathematician J. von Neumann, who was also involved in the atomic problem and asked the same question independently, submitted a paper three days later, on Monday 30 June 1941 (von Neumann 1941; see also von Neumann 1963). His solution complemented Taylor's solution – he noticed an energy integral for the set of ordinary differential equations and was able to obtain the solution in closed form. Later, the solution of this problem was published in the Soviet Union by L.I. Sedov (Sedov 1946, 1959), who also found the energy integral, and by other authors, R. Latter (1955) and J. Lockwood Taylor (1955).

We have seen that in obtaining the scaling law (0.6) and achieving the property of self-similarity an important role was played by dimensional analysis: the construction of dimensionless quantities from the arguments of the function

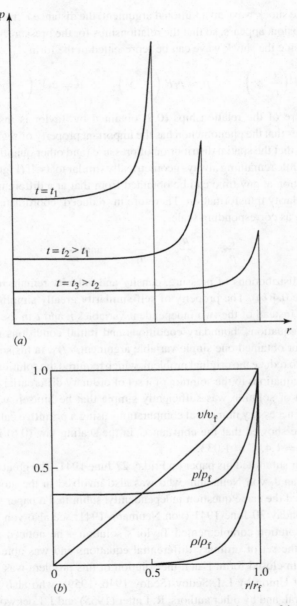

Figure 0.4. (*a*) Air pressure as a function of radius at various instants of time for the motion of air following an atomic explosion. The pressure distributions at various times are similar to one another. (*b*) Spatial distributions of the gas pressure, density and velocity in the reduced 'self-similar' coordinates ρ/ρ_f, p/p_f, u/u_f and r/r_f do not depend on time.

F with subsequent reduction in the number of arguments. The idea on which dimensional analysis is based is fundamental, but very simple: physical laws cannot depend on an arbitrary choice of basic units of measurement. The formal recipe for using dimensional analysis is very simple also. The main art, however, is not in using this simple tool but in finding, as G.I. Taylor did, the proper formulation or idealization of the problem in hand – an instantaneous concentrated very intense explosion in his case – that allows effective use of this tool. Here the key point is the concept of *intermediate asymptotics*: consideration of the phenomenon in intermediate time and space intervals.

It is important, however, to note that dimensional analysis is not always sufficient for obtaining self-similar solutions and scaling laws. Moreover, it can be claimed that as a rule it is not so and that the Taylor–von Neumann solution to the explosion problem was in fact a rare and lucky exception.

Here an instructive role is played by the paper by K.G. Guderley (1942) where, in a certain sense, the mirror image of the problem of a very intense explosion was considered. The formulation of this implosion problem is as follows.[1] On the wall of a spherical cavity of radius r_0 in an absolutely rigid vessel filled by gas of density ρ_0 (Figure 0.5) there is a uniform thin layer of a strong explosive. The latter is exploded instantaneously and uniformly over the wall and a strong spherical shock wave is formed. The shock wave converges to the center of the cavity. It is very intense, as in the case of a very intense explosion, so that the pressure behind the wave is much larger than the initial gas pressure p_0, which, as in the case of a very intense explosion, can be neglected. The shock wave comes to a focus at the center of the cavity at a time which we take as $t = 0$, so that the time before focusing will be negative, $t < 0$. Similarly to the case of an intense explosion, dimensional analysis gives for the radius of the shock wave

$$r_f = [E(-t)^2/\rho_0]^{1/5}\Phi(\eta, \gamma), \qquad \eta = \frac{r_0}{[E(-t)^2/\rho_0]^{1/5}} \qquad (0.10)$$

where as before E is the energy of the explosion and γ is the adiabatic index.

Seemingly the application of reasoning analogous to that for the case of an intense explosion would suggest that the argument η goes to infinity at $t \to 0$ and therefore can be neglected close to the focus, so that a formula analogous to (0.6) could be obtained:

$$r_f = C(\gamma)\left[\frac{E(-t)^2}{\rho_0}\right]^{1/5} \qquad (0.11)$$

---

[1] A detailed discussion of the Guderley problem can also be found in the books by Zeldovich and Raizer (1967) and Landau and Lifshitz (1987).

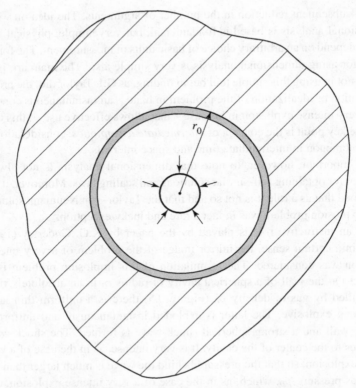

Figure 0.5. A very intense implosion in a spherical cavity. The explosive is placed on the wall of the cavity. The black dot shows the shock front as it comes to a focus at the centre of the cavity at $t = 0$.

In fact, this is not the case, for the following reason. In the case of implosion the function $\Phi(\eta, \gamma)$ at $\eta \to \infty$ does not tend to a finite non-zero limit as was the case for an explosion! However, it happens that at $\eta \to \infty$ the function $\Phi(\eta, \gamma)$ has a power-law-type behavior, $\Phi(\eta, \gamma) \sim C(\gamma)\eta^{-\beta}$ where $\beta = \beta(\gamma) = $ const > 0, so that at $t \to 0$, that is, close to the focus, the expression for the radius of the shock wave assumes the form

$$r_{\mathrm{f}} = C(\gamma)r_0^{-\beta}\left[\frac{E(-t)^2}{\rho_0}\right]^{\alpha/2} = A(-t)^{\alpha},$$

$$\alpha = \frac{2}{5}(1 + \beta), \qquad A = C(\gamma)r_0^{-\beta}\left(\frac{E}{\rho_0}\right)^{\alpha/2}. \qquad (0.12)$$

It is important to note that the exponent α cannot be obtained by dimensional analysis, as it was in the case of an intense explosion, but requires a more complicated technique, the solution of a *nonlinear eigenvalue problem*.

The Guderley (1942) solution as well as the solution to the 'impulsive load' problem which is in fact a one-dimensional analog of the implosion problem, obtained by von Weizsächer (1954) and Zeldovich (1956), introduced a new class of self-similar phenomena: *incomplete similarity and self-similar solutions of the second kind*. These problems are closely related to the concept of the renormalization group, well known in theoretical physics.

In what follows we will present in detail the ideas of dimensional analysis, physical similarity, self-similarity, intermediate asymptotics and the renormalization group. Our goal is to demonstrate in detail the many possibilities for application of these ideas and also the difficulties which can occur – throughout using many examples. Most of the examples in the present book are related to fluid dynamics: my experience shows that the elements of fluid mechanics are familiar to engineers, mathematicians and physicists. Those who are more interested in elasticity, fracture, fatigue or geophysical fluid dynamics can find additional examples in my book Barenblatt (1996). The examples ('Problems') considered in the present book should be considered as an essential part of the whole text.

Chapter 1
Dimensional analysis and physical similarity

1.1 Dimensions

1.1.1 Measurement of physical quantities, units of measurement. Systems of units

We say without any particular thought that the mass of water in a glass is 200 grams, the length of a ruler is 0.30 meters (12 inches), the half-life of radium is 1600 years, the speed of a car is 60 miles per hour. In general, we express all physical quantities in terms of numbers; these numbers are obtained by *measuring* the physical quantities. Measurement is the direct or indirect comparison of a certain quantity with an appropriate standard, or, to put it another way, with an appropriate *unit of measurement*. Thus, in the examples discussed above, the mass of water is compared with a *standard* – a unit of mass, the gram; the length of the ruler is compared with a unit of length, the meter; the half-lifetime of radium is compared with a unit of time, the year; and the velocity of the car is compared with a unit of velocity, the velocity of uniform motion in which a distance of one mile is traversed in a time equal to one hour.

The units for measuring physical quantities are divided into two categories: *fundamental units* and *derived units*. This means the following.

A class of phenomena (for example, mechanics, i.e. the motion and equilibrium of bodies) is singled out for study. Certain quantities are listed, and standard reference values – either natural or artificial – for these quantities are adopted as fundamental units; there is a certain amount of arbitrariness here. For example, when describing mechanical phenomena we may adopt mass, length and time standards as the fundamental units, though it is also possible to adopt other sets, such as force, length and time. However, these standards are insufficient for the description of, for example, heat transfer, and so the unit of temperature, the kelvin, is introduced. Additional standards also become necessary when one is studying electromagnetic phenomena, luminous phenomena or, indeed, subject areas quite outside the scope of physical science, such as economics.[1]

Once the fundamental units have been decided upon, derived units are obtained from the fundamental units using the definitions of the physical quantities involved. These definitions always involve describing at least a conceptual method for measuring the physical quantity in question. For example, velocity is by definition the ratio of the distance traversed during some interval of time to the size of that time interval. Therefore, the velocity of uniform motion in which one unit of length is traversed in one unit of time can be adopted as a unit of velocity. In exactly the same way, density is by definition the ratio of some mass to the volume occupied by that mass. Thus, the density of a homogeneous body that contains one unit of mass per unit of volume – a cube with a side equal to one unit of length – can be adopted as a unit of density, and so on. We see that it is precisely the class of phenomena under discussion, i.e., the complete set of physical quantities in which we are interested, which ultimately determines whether a given set of fundamental units is sufficient for its measurement. For example, it is impossible to define a unit for the measurement of density using only the fundamental units of length and time. It becomes possible to define such a unit by adding a unit of mass.

A set of fundamental units that is *sufficient* for measuring the properties of the class of phenomena under consideration is called a *system of units*. Until recently, the cgs (centimeter–gram–second) system, in which units for mass, length and time are used as the basic units and one gram[2] (g) is adopted as the unit of mass, one centimeter[3] (cm) as the unit of length and one second[4] (s) as the unit of time, has customarily been used.

The unit of velocity in this system is the velocity of uniform motion in which a distance of one centimeter is traversed in one second. This unit is written in the following way: cm/s. The unit of density in the cgs system is the density of a homogeneous body in which one cubic centimeter contains a mass of one

---

[1] Recently the analysis of economic and, especially, financial phenomena using the traditional approaches of applied mathematics has attracted serious attention. For such applications the correct definition and measurement of the quantities involved is of prime importance.
[2] The gram is one-thousandth of the mass of a specially fabricated standard mass, which is carefully preserved at the Bureau of Weights and Measures in Paris.
[3] The centimeter is one-hundredth of the length of a specially fabricated, carefully preserved standard length – the meter. There is another, more precise and universal definition of this standard based on a natural process: 1650 736.73 wavelengths *in vacuo* of the radiation corresponding to the transition between the $2p^{10}$ and $5d^5$ levels of the krypton-86 atom.
[4] The second is, by definition, 1/86 400 of a mean solar day. A more precise and universal definition of the second is 9192 621 770 periods of the radiation corresponding to the transition between two hyperfine levels in the ground state of the caesium-133 atom.

gram. This unit is written in the following way: g/cm³. This method of writing units is, to a certain extent, a matter of convention: for example, the ratio cm/s cannot be thought of as a quotient of the length standard – the centimeter – and the time standard – the second. Such a quotient would be totally meaningless: one may divide one number by another, but not an interval of length by an interval of time!

A system of units consisting of two units (a unit for the measurement of length and a unit for the measurement of time, for example the centimeter and the second) is sufficient for measuring the properties of *kinematic* phenomena, while a system of units consisting of only one length unit (for example the centimeter) is sufficient for measuring the properties of *geometric* objects.

However, in order to be able to measure the properties of *heat transfer*, the system of units for the measurement of mechanical quantities must be supplemented by an independent standard (the degree Kelvin (kelvin), a temperature standard, is convenient for this purpose). We would require an additional standard, for example a unit of electric current (the ampere) in order to be able to measure electromagnetic phenomena and so forth.

Note that a system of units need not be *minimal*, i.e. redundancy in its units need not be avoided. For example, one can use a system of units in which the unit of length is 1 cm, the unit of time is 1 s and the unit of velocity is 1 knot (approximately 50 cm/s). However, in this case, the velocity will not be numerically equal to the ratio of the distance traversed to the magnitude of the time interval in which the distance is traversed. We shall discuss this important point in greater detail below.

1.1.2 Classes of systems of units

Let us now consider, in addition to the cgs system, a second system, in which one kilometer (= 10⁵ cm) is used as the unit of length, one metric ton (= 10⁶ g) is used as the unit of mass and one hour (= 3600 s) is used as the unit of time. These two systems of units have the following property in common: standard quantities of the same physical nature (mass, length and time) are used as the fundamental units. Consequently, we say that these systems belong to the same *class*. To generalize, a set of systems of units that differ only in the magnitudes (but not in the physical nature) of the fundamental units is called a *class of systems of units*. The system just mentioned and the cgs system are members of the class in which standard lengths, masses and times are used as the fundamental units. If we choose to regard the cgs system as the *original system* in this class then the corresponding units for an arbitrary system in this class are as follows:

$$\text{unit of length} = \text{cm}/L,$$
$$\text{unit of mass} = \text{g}/M, \qquad (1.1)$$
$$\text{unit of time} = \text{s}/T,$$

where L, M and T are *positive numbers* that indicate the factors by which the fundamental units of length, mass and time *decrease* in passing from the original system (in this case, the cgs system) to another system in the same class. This class is called the *LMT* class.[5] The SI system has recently come into widespread use. This system, in which one meter ($= 100$ cm) is adopted as the unit of length, one kilogram ($= 1000$ g) as the unit of mass and one second as the unit of time, also belongs to the *LMT* class. Thus, when passing from the original system to the SI system, $M = 0.001$, $L = 0.01$ and $T = 1$.

Systems in the *LFT* class, where units for length, force and time are chosen as the fundamental units, are also frequently used. Using as original units 1 cm, 1 kgf and 1 s, the fundamental units for an arbitrary system in this class are as follows:

$$\text{unit of length} = \text{cm}/L,$$
$$\text{unit of force} = \text{kgf}/F, \qquad (1.2)$$
$$\text{unit of time} = \text{s}/T.$$

The unit of force in the original system, the kilogram-force (kgf), is the force that imparts an acceleration of 9.80665 m/s^2 to a mass equal to that of the standard kilogram.

We emphasize that a change in the magnitudes of the fundamental units in the original system of units does not change the class of systems of units. For example, a class in which the units of length, mass and time are given by

$$\frac{\text{m}}{L}, \qquad \frac{\text{kg}}{M}, \qquad \frac{\text{hr}}{T}$$

is the same as that defined in (1.1), *LMT*. The only difference is that the numbers L, M and T for a certain system of units (for example, the SI system) will be different for the two members, or *representations*, of the *LMT* class: in the second representation, we obviously have $L = 1$, $M = 1$ and $T = 3600$.

---

[5] The designation of a class of systems of units is obtained by writing down, in consecutive order, the symbols for the quantities whose units are adopted as the fundamental units. Such a symbol simultaneously denotes the *factor* by which the corresponding fundamental unit decreases upon passage from the original system to another system in the same class. The reader should be careful to distinguish between these two, closely related, meanings of L, M, T etc.

1.1.3 Dimension of a physical quantity

If the unit of length is decreased by a factor L and the unit of time is decreased by a factor T then the new unit of velocity is a factor LT^{-1} times smaller than the original unit, so that the numerical values of all velocities are increased by a factor LT^{-1}. Similarly, upon decreasing the unit of mass by a factor M and the unit of length by a factor L we find that the new unit of density is a factor $L^{-3}M$ smaller than the original unit, so that the numerical values of all densities are increased by a factor $L^{-3}M$. Other quantities may be treated similarly. The change in the numerical value of a physical quantity upon passage from one system of units to an arbitrary system within the same class is determined by its *dimension. The function that determines the factor by which the numerical value of a physical quantity changes upon passage from the original system of units to an arbitrary system within a given class is called the dimension function, or dimension,*[6] *of that quantity.* It is customary, following a suggestion of J.C. Maxwell, to denote the dimension of a quantity ϕ by $[\phi]$. We emphasize that the dimension function of a given physical quantity is determined for a specified class and is different in different classes of systems of units. For example, the dimension function of density ρ in the *LMT* class is $[\rho] = L^{-3}M$; in the *LFT* class it is $[\rho] = L^{-4}FT^2$.

Quantities whose numerical values are identical in all systems of units within a given class are called *dimensionless*; clearly, the dimension function is equal to unity for a dimensionless quantity. All other quantities are called *dimensional*.

We shall now cite a few additional examples. If (in the *LMT* class) the unit of length is decreased by a factor L, the unit of mass is decreased by a factor M and the unit of time is decreased by a factor T then the numerical values of all forces are increased by a factor LMT^{-2}. Indeed, according to Newton's second law, the net force f on a mass m is the product of the mass and its acceleration a:

$$f = ma.$$

For the decreases in the fundamental units mentioned at the start of this subsection, the numerical values of all masses are increased by a factor M and the numerical values of all accelerations are increased by a factor LT^{-2}. Now, *the dimensions of both sides of any equation with physical sense must be identical*: otherwise, an equality in one system of units would not be an equality in another

---

[6] Our use of the singular should be noted.

system, and this is not permissible for equations with physical sense.[7] Thus, we find that the dimension of force in the *LMT* class is

$$[f] = [m][a] = LMT^{-2}. \tag{1.3}$$

Analogously, the dimension of mass in the *LMT* class is M, while it is $[m] = L^{-1}FT^2$ in the *LFT* class; the dimension of energy, $[e]$, is L^2MT^{-2} in the *LMT* class and LF in the *LFT* class. In the *LMT* class, the ratio of velocity and distance divided by time is dimensionless. However, if we use the *LMTV* class, in which the unit of velocity (knot/V) is independent, this ratio has a dimension different from unity, $L^{-1}TV$. For instance, for a vessel travelling at 20 knots the ratio is equal to 20 if the unit of length is one nautical mile ($\sim$ 1850 meters) and equal to 37 if the unit of length is one kilometer, whereas the units of time and velocity, one hour and one knot respectively, are the same in each system.

Dimension functions possess two important properties, which we shall now discuss.

1.1.4 The dimension function is always a power-law monomial

We have seen that the dimension function is a power-law monomial in all the cases discussed above. This brings up the following question: are there physical quantities for which this is not so, and for which the dimensions in the *LMT* class are given, for example, by dimension functions of the form $L + M^2$, $e^L M$ or $\sin M \log T$? In fact, there are no such physical quantities, and the *dimension function for any physical quantity is always a power-law monomial*. This follows from a simple, naturally formulated (but actually very deep) physical principle: all systems within a given class are equivalent, i.e., there are no distinguished, somehow preferred, systems among them.

We shall prove this using the *LMT* class of systems; the reader may easily make the generalization to an arbitrary class of systems. By virtue of the fact that the systems within a given class are equivalent, the dimension in this class of any mechanical quantity a depends only on the ratios L, M and T (see subsection 1.1.3):

$$[a] = \phi(L, M, T). \tag{1.4}$$

---

[7] Equations which hold only in one system of units do exist and sometimes are very useful, although they have no physical sense. For instance, my colleague Professor A.Yu. Ishlinsky proposed a formula for the time taken to drive a given distance in Moscow: the time in minutes is equal to the distance in kilometers plus the number of traffic lights. Of course, the formula time = distance + number of traffic lights does not work in other units, and therefore has no physical sense.

If there existed some *distinguished system* within the *LMT* class, it would be necessary to include in (1.4) the relationship between the system of units we are working in and the distinguished system. In this case, the dimension function ϕ would depend on three additional arguments, ℓ_0/ℓ_d, m_0/m_d and t_0/t_d, the ratios of the units of length, mass and time, ℓ_0, m_0 and t_0, in the original system of the *LMT* class and the corresponding units, ℓ_d, m_d and t_d, in the distinguished system. According to the equivalence principle formulated above, this cannot be so: the dimension function ϕ depends only upon the dimensions L, M and T in the *LMT* class, independently of which system is adopted as the original system.

To continue our proof, we shall now choose two systems of units within the *LMT* class: system 1, which is obtained from the original system upon decreasing the fundamental units by factors of L_1, M_1 and T_1, and system 2, which is obtained from the original system upon decreasing the fundamental units by factors of L_2, M_2 and T_2.

By the definition of dimension, the numerical value of the quantity under discussion, equal, say, to a in the original system, is $a_1 = a\phi(L_1, M_1, T_1)$ in the first system, and $a_2 = a\phi(L_2, M_2, T_2)$ in the second system. Thus, we have

$$\frac{a_2}{a_1} = \frac{\phi(L_2, M_2, T_2)}{\phi(L_1, M_1, T_1)}. \tag{1.5}$$

We now note that by virtue of the equivalence of systems within a given class, we may assume that system 1 is the original system of the class, without altering the class. In this case, system 2 can be obtained from the new original system (system 1) by decreasing the fundamental units by factors of L_2/L_1, M_2/M_1 and T_2/T_1, respectively. Consequently, the numerical value a_2 of the quantity under discussion in the second system of units, is, by the definition of the dimension function,

$$a_2 = a_1\phi(L_2/L_1, \ M_2/M_1, \ T_2/T_1);$$

we emphasize that a_1, the numerical value of the quantity a in system 1, remains unchanged under the change in original system made above. Thus $a_2/a_1 = \phi(L_2/L_1, \ M_2/M_1, \ T_2/T_1)$. Setting this expression equal to that in (1.5), we obtain the following equation for the dimension function ϕ:

$$\frac{\phi(L_2, M_2, T_2)}{\phi(L_1, M_1, T_1)} = \phi(L_2/L_1, \ M_2/M_1, \ T_2/T_1). \tag{1.6}$$

Equations of this type are called functional equations. We shall now show that only power-law monomials satisfy this equation.

To solve (1.6), we differentiate[8] both sides of this equation with respect to L_2 and *then* set $L_2 = L_1 = L$, $M_2 = M_1 = M$ and $T_2 = T_1 = T$. We find that

$$\frac{\partial_L \phi(L, M, T)}{\phi(L, M, T)} = \frac{1}{L}\partial_L \phi(1, 1, 1) = \frac{\alpha}{L}, \tag{1.7}$$

where the quantity $\alpha = \partial_L \phi(1, 1, 1)$ is a *constant independent of L, M and T*. Integrating (1.7), we find that

$$\phi(L, M, T) = L^\alpha C_1(M, T). \tag{1.8}$$

Substituting this expression into (1.6), we obtain an equation for the function C_1 of the same form as (1.6) but with one argument fewer:

$$\frac{C_1(M_2, T_2)}{C_1(M_1, T_1)} = C_1(M_2/M_1, T_2/T_1). \tag{1.9}$$

Once again, we proceed in the same way: we differentiate both sides of (1.9) with respect to M_2 and set $M_2 = M_1 = M$ and $T_2 = T_1 = T$:

$$\frac{\partial_M C_1(M, T)}{C_1(M, T)} = \frac{1}{M}\partial_M C_1(1, 1) = \frac{\beta}{M},$$

from which

$$C_1 = M^\beta C_2(T), \tag{1.10}$$

where $\beta = \partial_M C_1(1, 1)$ is a constant similar to α. Following the same line of reasoning again, we find that

$$C_2(T) = C_3 T^\gamma,$$

so that

$$\phi = C_3 L^\alpha M^\beta T^\gamma. \tag{1.11}$$

The constant C_3 is obviously equal to unity, since $L = M = T = 1$ means that the fundamental units remain unchanged, so that the value of the quantity *a* must remain unchanged and $\phi(1, 1, 1) = 1$.

So, we have shown that the solution to the functional equation (1.6) is the power-law monomial $L^\alpha M^\beta T^\gamma$, where α, β and γ are constants; therefore the dimension of any mechanical quantity and, by extension, any other physical quantity can be expressed in terms of a power-law monomial.

Let us look at what would happen if, for instance, the unit of length were a distinguished unit, equal, say, to $\ell_d = 1$ foot. (Originally, it was taken as

---

[8] It is natural to assume that the dimension function is smooth, although, in fact, only the assumption of continuity is enough.

the length of the foot of an English king.) In this case the ratio ℓ_0/ℓ_d of the fundamental unit of length in the original system ℓ_0 to the foot, i.e. the length of the former in feet, will be significant and should be included in the arguments of the dimension function. Therefore, relation (1.6) would be of the form

$$\frac{\phi(L_2, M_2, T_2, \ell_0/\ell_d)}{\phi(L_1, M_1, T_1, \ell_0/\ell_d)} = \phi\left(\frac{L_2}{L_1}, \frac{M_2}{M_1}, \frac{T_2}{T_1}, \frac{\ell_0/\ell_d}{L_1}\right).$$

Differentiating by L_2 and then setting $L_2 = L_1 = L$, $M_2 = M_1 = M$ and $T_2 = T_1 = T$, we obtain

$$\frac{\partial_L\phi(L, M, T)}{\phi(L, M, T)} = \frac{1}{L}\,\partial_L\phi\left(1, 1, 1, \frac{\ell_0/\ell_d}{L}\right) \neq \frac{\text{const}}{L}.$$

Thus, if we give up the principle that all systems of units within a given class are equivalent, i.e. that there is no distinguished system in the class, the main result of this principle – that dimension functions are power monomials – does not hold.

It should be noted that systems of units convenient for use with some special classes of problem have frequently been proposed. For example, Kapitza (1966) proposed a natural system of units for classical electrodynamics. Kapitza's system uses the classical radius of the electron as the unit of length, the rest-mass energy of the electron as the unit of energy and the mass of the electron as the unit of mass. This system is convenient in classical electrodynamics problems, since it allows one to avoid very large or very small numerical values for all quantities of practical interest. It is important to note that Kapitza's system is not 'distinguished' in the sense described above: the dimensions of physical quantities for an arbitrary system in the *LEM* class (E is the symbol for energy) do not depend on the ratios of the units of length, energy and mass in an original system in the class to the units in Kapitza's system.

1.1.5 Quantities with independent dimensions

The quantities $a_1, \ldots, a_k$ are said to have *independent dimensions* if none of these quantities has a dimension function that can be represented as a product of the powers of the dimensions of the remaining quantities.

For example, density ($[\rho] = LM^{-3}$), velocity ($[U] = LT^{-1}$) and force ($[f] = LMT^{-2}$), have independent dimensions. To show this, let us assume that, on the contrary, only two of the three have independent dimensions. Then, since the dimension functions for both density and force contain M and the dimension function for velocity does not, there must exist numbers x and y such that $[f] = [\rho]^x[U]^y$. Substituting the expressions for the dimensions $[f]$, $[\rho]$ and

$[U]$ in terms of L, M and T into this relation, we find that

$$LMT^{-2} = (ML^{-3})^x (LT^{-1})^y. \tag{1.12}$$

Equating the exponents of L, M and T on the two sides of the equation, we obtain a system of three equations for the two unknowns x and y:

$$-3x + y = 1, \qquad x = 1, \qquad y = 2, \tag{1.13}$$

which obviously has no solution; $x = 1$ and $y = 2$ do not satisfy the first equation. So, we come to a contradiction, and we conclude that our assumption was false. In fact, it is easy to see that the dimensions of density, velocity and pressure are dependent: the dimension of pressure (force per unit area), $L^{-1}MT^{-2}$, is equal to the product of the dimension of density and the square of the dimension of velocity.

Furthermore, it is clear that none of the quantities $a_1, \ldots, a_k$ having independent dimensions can be dimensionless: the dimension of a dimensionless quantity, which is equal to unity, is equal to the product of the dimensions of the remaining quantities (whatever they are) raised to the power zero.

The fact which will be important below is that *it is always possible to pass from a chosen original system of units to some other system, within the same class, such that any quantity, say a_1, in the set of quantities with independent dimensions $a_1, \ldots, a_k$ changes its numerical value by a specified factor A_1 while the other quantities remain unchanged.*

Problem *Prove the above-mentioned property.*

Solution. Passing, in a given class of systems of units $PQ \ldots (P, Q, \ldots$ denote the symbols L, M, T and/or other similar quantities), from a chosen original system to an arbitrary one we obtain new values $a'_1, \ldots, a'_k$ of the parameters $a_1, \ldots, a_k$:

$$a'_1 = a_1 P^{\alpha_1} Q^{\beta_1} \cdots, \qquad a'_2 = a_2 P^{\alpha_2} Q^{\beta_2} \cdots, \qquad \ldots,$$
$$a'_k = a_k P^{\alpha_k} Q^{\beta_k} \cdots, \tag{1.14}$$

where the powers $\alpha_1, \beta_1, \ldots, \alpha_k, \beta_k$ are determined by the dimensions of $a_1, \ldots, a_k$, respectively. We want to find the system such that

$$a'_1 = A_1 a_1, \qquad a'_2 = a_2, \qquad \ldots, \qquad a'_k = a_k.$$

Therefore, for $P, Q, \ldots$ a system of equations is obtained:

$$P^{\alpha_1} Q^{\beta_1} \cdots = A_1, \qquad P^{\alpha_2} Q^{\beta_2} \cdots = 1, \qquad P^{\alpha_k} Q^{\beta_k} \cdots = 1. \tag{1.15}$$

Taking logarithms, we obtain a system of linear equations:

$$\alpha_1 \ln P + \beta_1 \ln Q + \cdots = \ln A_1,$$
$$\alpha_2 \ln P + \beta_2 \ln Q + \cdots = 0,$$
$$\vdots$$
$$\alpha_k \ln P + \beta_k \ln Q + \cdots = 0. \qquad (1.16)$$

This system has at least one solution. Indeed, it is insoluble only if the left-hand side of the first equation is a linear combination of the left-hand sides of the remaining equations,

$$\alpha_1 \ln P + \beta_1 \ln Q + \cdots = c_2(\alpha_2 \ln P + \beta_2 \ln Q + \cdots) + \cdots$$
$$+ c_k(\alpha_k \ln P + \beta_k \ln Q + \cdots) \qquad (1.17)$$

where $c_2, \ldots, c_k$ are constants. This would imply, if we return to the exponents from the logarithms, that

$$P^{\alpha_1} Q^{\beta_1} \cdots = (P^{\alpha_2} Q^{\beta_2} \cdots)^{c_2} \cdots (P^{\alpha_k} Q^{\beta_k} \cdots)^{c_k},$$

giving

$$[a_1] = [a_2]^{c_2} \cdots [a_k]^{c_k} \qquad (1.18)$$

so that the dimension of a_1 would be equal to the product of the powers of the dimensions of $a_2, \ldots, a_k$, which would contradict the assumption that the dimensions of the quantities $a_1, \ldots, a_k$ are independent. Thus the property is proved.

1.2 Dimensional analysis

1.2.1 Governing parameters

In any physical study (theoretical or experimental), we attempt to obtain relationships among the quantities that characterize the phenomenon being studied. Thus, the problem always reduces to determining one or several relationships of the form

$$a = f(a_1, \ldots, a_k, b_1, \ldots, b_m), \qquad (1.19)$$

where a is the quantity being determined in the study, and its $n = k + m$ arguments $a_1, \ldots, a_k, b_1, \ldots, b_m$ are assumed to be given; they are called *governing parameters*. The governing parameters in (1.19) are divided up in such a way that the k parameters $a_1, \ldots, a_k$ have independent dimensions while the dimensions of the m parameters $b_1, \ldots, b_m$ can be expressed as products of powers of the dimensions of the parameters $a_1, \ldots, a_k$:

$$[b_1] = [a_1]^{p_1} \cdots [a_k]^{r_1},$$

$$\vdots$$

$$[b_i] = [a_1]^{p_i} \cdots [a_k]^{r_i}, \tag{1.20}$$

$$\vdots$$

$$[b_m] = [a_1]^{p_m} \cdots [a_k]^{r_m}.$$

Such a division may always be made. In some special cases, we might have $m = 0$ (if the dimensions of all the governing parameters are independent) or $k = 0$ (if all the governing parameters are dimensionless). In general $k > 0$, $m > 0$.

The dimension of the quantity a to be determined must be expressible in terms of the dimensions of the governing parameters in the first group, $a_1, \ldots, a_k$:

$$[a] = [a_1]^p \cdots [a_k]^r. \tag{1.21}$$

If this were not so, the dimensions of the quantities $a, a_1, \ldots, a_k$ would be independent. Then, by the property proved in subsection 1.1.5, it would be possible to change the value of the quantity a by an arbitrary factor, via a change in the system of units within the class in question, and leave the quantities $a_1, \ldots, a_k$ unchanged. In doing so, the quantities $b_1, \ldots, b_m$, whose dimensions can be expressed in terms of the dimensions of the quantities $a_1, \ldots, a_k$, would likewise remain unchanged. Thus, the quantity to be determined, a, could be changed by any amount while the values of all the governing parameters remained unchanged; this is impossible if the list of governing parameters is complete. Thus, there always exist numbers $p, \ldots, r$ such that (1.21) holds.

1.2.2 Transformation to dimensionless parameters. Generalized homogeneity. Π-theorem

We shall now introduce the parameters

$$\Pi = \frac{a}{a_1^p \cdots a_k^r}$$

$$\Pi_1 = \frac{b_1}{a_1^{p_1} \cdots a_k^{r_1}}, \quad \cdots \quad \Pi_i = \frac{b_i}{a_1^{p_i} \cdots a_k^{r_i}}, \quad \cdots$$

$$\Pi_m = \frac{b_m}{a_1^{p_m} \cdots a_k^{r_m}}, \tag{1.22}$$

where the exponents of the governing parameters with independent dimensions are chosen such that all the parameters $\Pi, \Pi_1, \ldots, \Pi_m$ are dimensionless. Relation (1.19) may be rewritten, replacing the parameters $a, b_1, \ldots, b_m$ (whose dimensions depend on those of the parameters $a_1, \ldots, a_k$) by the dimensionless quantities $\Pi, \Pi_1, \ldots, \Pi_m$ defined in (1.22) and keeping the parameters $a_1, \ldots, a_k$. We find that

$$\Pi = \frac{f(a_1, \ldots a_k, b_1, \ldots, b_m)}{a_1^p \cdots a_k^r}$$

$$= \frac{1}{a_1^p \cdots a_k^r} f\left(a_1, \ldots a_k, \Pi_1 a_1^{p_1} \cdots a_k^{r_1}, \ldots, \Pi_m a_1^{p_m} \cdots a_k^{r_m}\right).$$

Thus, we find that

$$\Pi = \mathcal{F}(a_1, \ldots a_k, \Pi_1, \ldots, \Pi_m), \tag{1.23}$$

where $\mathcal{F}$ is a certain function.

Now, the most important point to be discussed here is as follows. We have already seen that it is always possible to pass to a system of units within the class in question such that any one of the parameters with independent dimensions $a_1, \ldots, a_k$, let us say a_1, is changed by an arbitrary factor, the remaining parameters, $a_2, \ldots, a_k$, remaining unchanged. Obviously, the dimensionless arguments $\Pi_1, \ldots, \Pi_m$ of the function $\mathcal{F}$ and the value of the dimensionless function Π also remain unchanged under such a transformation. It follows from this that the function $\mathcal{F}$ is in fact independent of the argument a_1. In exactly the same way, it can be shown that it is also independent of the arguments $a_2, \ldots, a_k$, so that $\mathcal{F} = \Phi(\Pi_1, \ldots, \Pi_m)$. Equation (1.23) can therefore in fact be written in terms of a function Φ of m rather than $n = k + m$ arguments:

$$\Pi = \Phi(\Pi_1, \ldots, \Pi_m). \tag{1.24}$$

However, since $\Pi = f/a_1^p \cdots a_k^r$, it follows that *any function f that defines some physical relationship possesses the property of a generalized homogeneity or symmetry*, i.e. it can be written in terms of a function of a smaller number of variables and is of the following special form:

$$f(a_1, \ldots a_k, b_1, \ldots, b_m) = a_1^p \cdots a_k^r \Phi\left(\frac{b_1}{a_1^{p_1} \cdots a_k^{r_1}}, \ldots, \frac{b_m}{a_1^{p_m} \cdots a_k^{r_m}}\right). \tag{1.25}$$

These results lead to the central result in dimensional analysis, the so-called Π-theorem: *a physical relationship between some dimensional (generally speaking) quantity and several dimensional governing parameters can be rewritten as a relationship between a dimensionless parameter and several dimensionless products of the powers of governing parameters; the number of dimensionless products is equal to the total number of governing parameters minus the number of governing parameters with independent dimensions.* The term 'physical relationship' is used to emphasize that it should be valid in all systems of units.

Note that the Π-theorem is, in fact, obvious at an intuitive level. Indeed, it is clear that physical laws cannot depend on the choice of units. Therefore, it must be possible to express them using relationships between quantities that do not depend on this arbitrary choice, i.e., dimensionless combinations of the variables. This was realized long ago, and concepts from dimensional analysis were in use long before the Π-theorem had been explicitly recognized, formulated and proved formally. The outstanding names that should be mentioned here are Galilei, Newton, Fourier, Maxwell, Reynolds and Rayleigh.

Dimensional analysis may be successfully applied (see below) in theoretical studies where a mathematical model of the problem is available, in the processing of experimental data and also in the preliminary analysis of physical phenomena preceding the construction of each model. The point that we are trying to make here is the following.

In order to determine the functional dependence of some quantity a, (1.19), on each of the governing parameters, it is necessary to either measure or calculate the function f for, let us say, 10 values of each governing parameter. Of course, the number 10 is somewhat arbitrary; a smaller number of measurements or calculations may suffice for some smooth functions, while even 100 measurements are insufficient for other functions. Thus, it is necessary to carry out a total of 10^{k+m} measurements or calculations to determine a. After applying dimensional analysis, the problem is reduced to one of determining a function Φ of m dimensionless arguments $\Pi_1, \ldots, \Pi_m$, and only 10^m (i.e. a factor of 10^k fewer) experiments or calculations are required to determine this function. As a result, we reach the following important conclusion: *the amount of work required to determine the desired function is reduced by as many orders of magnitude as there are governing parameters with independent dimensions.*

The following question naturally arises: if such substantial advantages are obtained for $n = k, m = 0$, why not go to a class of systems of units in which the dimensions of all the quantities $a_1, \ldots, a_k, b_1, \ldots, b_m$ are independent?

Actually, nothing is gained in general by transforming to such a class. We will show this using as an example a problem where quantities with dimensions of length ℓ, time τ and velocity v are included among the governing parameters. We will then change to the *LTV* class of systems, where the unit of velocity is independent. However, without modification the formula $v = s/t$ (where s is the distance travelled, and t is the time of travel) is not valid in

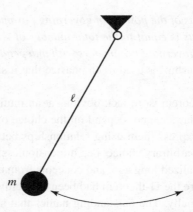

Figure 1.1. A pendulum performs small oscillations. Experiment shows that the period of small oscillations is independent (Galilei) of the maximum deviation of the pendulum.

this class; it must be replaced by the formula $v = As/t$, where A is a constant having dimension $L^{-1}TV$ (see subsection 1.1.3). In general, therefore, the quantity A must also be included among the governing parameters, thereby increasing the number of governing parameters by one. And, in general, the difference $m = n - k$ between the total number of governing parameters and the number of governing parameters with independent dimensions remains unchanged; thus, generally speaking, there is no advantage in transforming to a new class of systems of units. However, in some special cases it may turn out that the additional parameters, as is the case for A, happen to be non-essential. In such cases, transforming to a new class increases the number of parameters with independent dimensions and so is useful. We will see examples of this below.

1.2.3 Problems

Problem 1. *Derive, using dimensional analysis, the formula for the period θ of small oscillations of a pendulum,*

$$\theta = 2\pi\sqrt{\frac{\ell}{g}} \approx 6.28\sqrt{\frac{\ell}{g}}. \tag{1.26}$$

Here ℓ is the length of the pendulum (Figure 1.1), and g is the gravitational acceleration.

Solution: It is arguable that in principle the period θ depends upon the following governing parameters:

1. the length of the pendulum ℓ;
2. the mass of the bob m;
3. the gravitational acceleration g: if there were no gravity then the pendulum would not oscillate.

The dimensions of the quantities involved are as follows:

$$[\theta] = T, \qquad [\ell] = L, \qquad [m] = M, \qquad [g] = LT^{-2}. \qquad (1.27)$$

The dimensions of the governing parameters, ℓ, m, g, are independent (each of them contains a dimension absent in the others). Therefore, using the notation defined at (1.19), $k = n = 3$. It is easy to show that $[\theta] = [\ell]^{1/2}[g]^{-1/2}$; then, from (1.22) we obtain

$$\Pi = \frac{\theta}{\ell^{1/2}[g]^{-1/2}}. \qquad (1.28)$$

In this case, $m = n - k = 0$, so that there are no parameters Π_i, and the function Φ in (1.24) and (1.25) is a constant. Therefore

$$\theta = \text{const}\sqrt{\frac{\ell}{g}}. \qquad (1.29)$$

The constant in (1.29) can be determined fairly accurately from a single experiment, which the reader may carry out by measuring the period of oscillation of a weight hung on a thread. With this step the derivation of formula (1.26) will be complete. This derivation (which is due to the French mathematician P. Appell) is instructive. It would seem that we have succeeded in obtaining an answer to an interesting problem from nothing – except a list of the quantities on which the period of oscillation of the pendulum is expected to depend, and a comparison (analysis) of their dimensions. In fact, this is not completely true: under this argument lies a deep physical model – idealization, like in the problem of G.I. Taylor considered in the introduction, – and observation: the amplitude-independence of the period for small oscillations, the possibility of neglecting the decay of oscillations due to the drag of ambient air etc.

Problem 2. *Prove, using dimensional analysis, Pythagoras' theorem (see also Migdal 1977)*

$$c^2 = a^2 + b^2. \qquad (1.30)$$

Solution: Consider Figure 1.2. The area S_c of the largest right-angled triangle is determined by its hypotenuse c and, for definiteness, the smaller of its acute

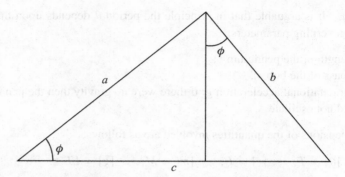

Figure 1.2. A proof of Pythagoras' theorem using dimensional analysis.

angles ϕ. Thus $S_c = f(c, \phi)$, and so $k = 1$, $m = 1$ since the only parameter with an independent dimension is c. Also, $\Pi = S_c/c^2$. We have

$$\Pi = \Phi(\Pi_1), \qquad \Pi_1 = \phi, \qquad S_c = c^2\Phi(\phi). \tag{1.31}$$

The perpendicular to the hypotenuse of the basic triangle divides it into two geometrically similar right triangles with hypotenuses equal to a and b. Equation (1.31), obtained by dimensional analysis, yields for the areas of these two triangles

$$S_a = a^2\Phi(\phi), \qquad S_b = b^2\Phi(\phi), \tag{1.32}$$

where the function Φ is the same as for the larger triangle. The sum of the areas of the two triangles S_a and S_b is equal to the area of the larger triangle S_c:

$$S_c = S_a + S_b, \qquad c^2\Phi(\phi) = a^2\Phi(\phi) + b^2\Phi(\phi). \tag{1.33}$$

Cancelling $\Phi(\phi)$ in the latter equation we obtain the desired result (1.30).

Here, there is an essential assumption underlying the simple procedure of dimensional analysis: the Euclidean nature of geometry. In both the Riemann and Lobachevskii geometries, however, there is an intrinsic parameter λ with the dimension of length, and the function Φ will depend on two arguments. Therefore instead of the relations (1.31), (1.32) different relations will be obtained:

$$S_c = c^2\Phi\left(\phi, \frac{\lambda}{c}\right), \qquad S_a = a^2\Phi\left(\phi, \frac{\lambda}{a}\right), \qquad S_b = b^2\Phi\left(\phi, \frac{\lambda}{b}\right), \tag{1.34}$$

and it is impossible to cancel Φ in relation (1.33), so the proof presented above no longer holds: Pythagoras' theorem in non-Euclidean geometries is not valid.

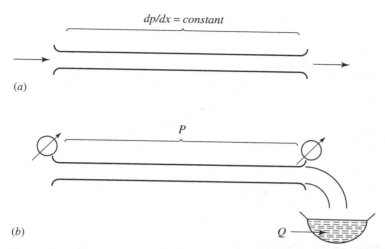

Figure 1.3. (*a*) A schematic diagram showing the experiments of Reynolds. The pipe is long, and the intermediate part, where the pressure gradient dp/dx is constant, occupies a major part of the pipe and determines the drag. (*b*) A schematic diagram showing the experiments of Bose, Rauert and Bose. The time τ required to fill a vessel of volume Q and the pressure drop between the ends of the pipe P were measured for the steady flow of various fluids through the pipe.

Problem 3. *Derive, using dimensional analysis, the relation for the hydrodynamic drag in a long cylindrical pipeline* (Figure 1.3(*a*)).

Solution: The basic assumption is that the drag in a long cylindrical pipeline is determined by the part intermediate between the pipe entrance and exit. In this part, as is confirmed by experiment, the pressure gradient is constant: it does not depend on the position x of the cross-section. Therefore

$$\frac{dp}{dx} = f(U, D, \rho, \mu), \tag{1.35}$$

where U is the velocity averaged over the cross-section, i.e. the total discharge rate divided by the cross-sectional area, D is the pipe diameter and ρ and μ are the fluid properties, its density and dynamic viscosity. The dimensions of these quantities in the *LMT* class are as follows:

$$\left[\frac{dp}{dx}\right] = \frac{M}{L^2T^2}, \quad [U] = \frac{L}{T}, \quad [D] = L, \quad [\rho] = \frac{M}{L^3}, \quad [\mu] = \frac{M}{LT}. \tag{1.36}$$

(The dimension of the viscosity, μ, can be obtained easily from the relation $\tau = \mu V/h$; this relation expresses the shear stress τ in Newton's experiment

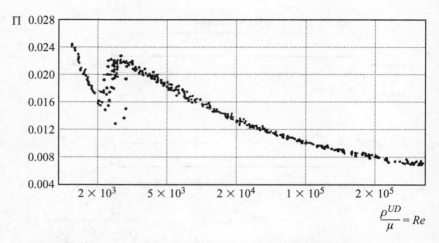

Figure 1.4. The dimensionless pressure drop per unit length of pipe, Π, of fluid passing through a pipe as a function of Reynolds parameter $Re = \rho UD/\mu$. With the exception of the transition region between laminar and turbulent flow, the data from different experiments all lie on a single curve. The complicated nature of the curve indicates that the flow regime changes as a function of Re, which is the only parameter that determines the global structure of the flow.

with a fluid layer of thickness h between two plates, one of which moves with velocity V and the second of which is at rest. The shear stress has the dimension of pressure, force per unit area). Thus $k = 3$, $m = 1$: the dimensions of U, D and ρ are obviously independent. The dimensions of dp/dx and μ are

$$\left[\frac{dp}{dx}\right] = [U]^2[D]^{-1}[\rho], \quad \text{and} \quad [\mu] = [U][D][\rho], \quad \text{giving}$$

$$\Pi = \frac{dp/dx}{U^2 D^{-1}\rho}, \qquad \Pi_1 = \frac{\mu}{UD\rho} \tag{1.37}$$

Thus, for this case, relation (1.25) takes the form

$$\frac{dp}{dx} = f(U, D, \rho, \mu) = U^2 D^{-1}\rho\,\Phi\left(\frac{\mu}{UD\rho}\right), \tag{1.38}$$

so that the function f of four variables is expressed via a function of one single variable, $\rho UD/\mu$.

This argument was performed for the first time by the English fluid mechanist Osborne Reynolds at the end of the nineteenth century. He examined the experimental data on flow in pipes available at the time and found that, to quite good accuracy, the experimental data in the coordinates $\rho UD/\mu$ and Π lay on a single curve (see Figure 1.4, where this relationship is illustrated using more recent data). This was a great success and, following a proposal by the prominent physicist A. Sommerfeld, the dimensionless parameter $\rho UD/\mu$ was later named the Reynolds number in honour of its originator.

Figure 1.4 shows clearly that the flow regimes at small and large values of the Reynolds number $Re = \rho UD/\mu$ are different. Nowadays we know why – the flow at small values of the Reynolds number is laminar whereas the flow at large values is turbulent. The experimental data in the transition region show strong scatter; they do not follow a single line. This indicates that the basic model leading to the relation (1.35) is invalid in this region and that some additional factors should be taken into account.

Problem 4. *Subject to dimensional analysis the results of different experiments for turbulent flow in pipes.*

Many years after the work of Reynolds mentioned above, the physico-chemists E. Bose, D. Rauert and M. Bose published a series of experimental studies of internal turbulent friction in various fluids (Bose and Rauert 1909; Bose and Bose 1911). The experiments were carried out in the following way (Figure 1.3(b)).

Various fluids, water, chloroform, bromoform, mercury, ethyl alcohol, etc., were allowed to flow through a pipe (not necessarily a long one) in a regime of steady turbulence. The time τ required to fill a bowl with a certain fixed volume Q and the pressure drop P between the ends of the pipe were measured. As was customary, the results of the measurements were represented in the form of a series of tables and curves (similar to those in Figure 1.5) showing P as a function of τ.

It was Th. von Kármán, at the time a young researcher, who later became one of the greatest applied mathematicians, who subjected the results of Bose, Bose and Rauert to a processing procedure using what is now called dimensional analysis (see von Kármán 1957).

Solution. The pressure drop between the ends of the pipe, P, depends on the time τ required for the bowl to be filled and on its volume Q as well as on the properties of the fluid, its viscosity μ and density ρ. It is instructive that von Kármán retained the none-too-promising original parameters Q and τ chosen by the experimenters. The dimensions of the quantities in question, for definiteness, in the LMT class, are given by the following expressions:

$$[P] = \frac{M}{LT^2}, \qquad [\tau] = T, \qquad [Q] = L^3, \qquad [\mu] = \frac{M}{LT}, \qquad [\rho] = \frac{M}{L^3}.$$

The first three governing parameters, τ, Q and μ, have independent dimensions; the dimensional formula for μ contains the mass while those for the other two

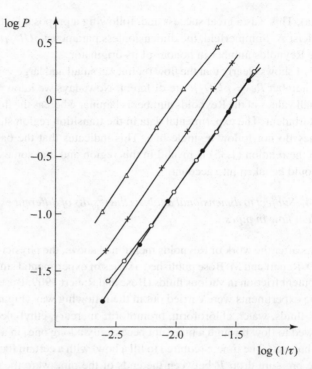

Figure 1.5. The experimental results of Bose, Rauert and Bose in their original form: O, water; ●, chloroform; +, bromoform; △, mercury (P in kgf/cm^2 and τ in seconds). The curves are different for the different fluids. From von Kármán (1957).

governing parameters do not. The dimension of μ therefore cannot appear in the dimensions of the other governing parameters with any exponent other than zero. Furthermore, the dimensional formula for Q contains L alone, and the dimensional formula for τ contains T alone. It is therefore impossible to obtain the dimension of any one of these quantities in terms of the dimensions of the other two. However, the dimension of the parameter ρ can be expressed as a product of the dimensions of τ, Q and μ raised to various powers:

$$[\rho] = [\tau][Q]^{-2/3}[\mu].$$

The dimension of the pressure drop P can also be expressed in terms of the dimensions of the governing parameters τ, Q and μ:

$$[P] = [\tau]^{-1}[Q]^{0}[\mu].$$

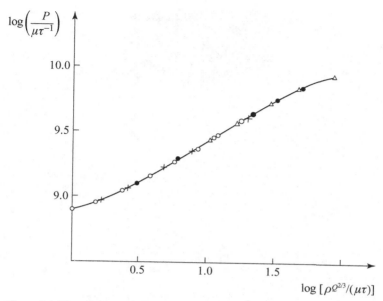

Figure 1.6. The experimental results of Bose, Rauert and Bose as represented by von Kármán, who used dimensional analysis. All the experimental points lie on a single curve. From von Kármán (1957).

Thus, (1.19), $k = 3$, $m = 1$. Dimensional analysis yields

$$\Pi = \Phi(\Pi_1), \qquad (1.39)$$

where

$$\Pi = \frac{P}{\mu \tau^{-1}}, \qquad \Pi_1 = \frac{\rho}{\mu \tau Q^{-2/3}}.$$

Therefore the search for the desired relationship between the pressure drop and the four parameters that govern it, $P = f(\tau, Q, \mu, \rho)$, reduces to the determination of a single function Φ of one composite parameter, the function $\Phi(\Pi_1)$, since equation (1.39) can be written in the form

$$P = \frac{\mu}{\tau} \Phi \left(\frac{\rho}{\mu \tau Q^{-2/3}} \right).$$

This means that all the experimental points should lie along a single curve in the coordinates $\rho/(\mu \tau Q^{-2/3})$, $P/(\mu \tau^{-1})$. Von Kármán's processing of the measured data of E. Bose, Rauert and M. Bose confirmed this (Figure 1.6). It is clear that if dimensional analysis had been carried out beforehand, the amount of

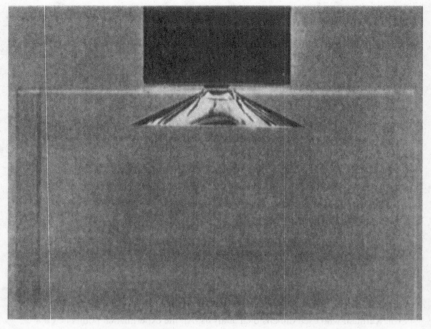

Figure 1.7. When a punch is pressed into a block of fused silica, a conical crack is formed (Benbow 1960).

experimental work required of the physico-chemists would have been reduced by a large factor.

Problem 5. *Derive, using dimensional analysis, the relation between the diameter of the conical crack formed in a brittle block under a punch and the applied load.*

In the remarkable experiments of Roesler (1956) and Benbow (1960) a punch with a small flat point was pressed slowly into the face of a cubic block of transparent brittle material (fused silica, see Figure 1.7). A perfect conical crack was formed under the punch and, as the load increased, the crack increased in size. The diameter of the base of the conical crack rapidly became much larger than the diameter of the flat point of the punch.

Solution.   The conical crack under the punch is in a state of *mobile equilibrium*: any increase in the load leads to its extension. It is known from the theory of elasticity (see Broberg 1999) that the stress directly underneath the crack tip,

σ, decreases in inverse proportion to the square root of the distance s from the crack tip:

$$\sigma \sim \frac{N}{\sqrt{s}},$$

where N is a constant. For a crack in a state of mobile equilibrium the principle of autonomy, i.e. universality, is valid (Barenblatt 1959a, 1962; Broberg 1999), according to which the elastic field near all crack tips in a given material under identical external conditions is identical. From this principle it follows that the *stress intensity factor* N is determined solely by the value of a constant of the material:

$$N = \frac{K}{\pi}.$$

The quantity K is called the *cohesion modulus* or *fracture toughness*; the factor π is included for historical reasons.

From the above considerations, it is natural to assume that the diameter D of the base of the conical crack depends on the load P and on the properties of the material, its cohesion modulus or fracture toughness K and Poisson ratio v. (Young's modulus E does not enter the set of governing parameters because the loads are prescribed but not the displacements.) For an intermediate stage when the loads are sufficiently large but not too large, the diameter of the punch d is much smaller than the diameter D of the base of the cone and the size Δ of the block is much larger than the diameter of the base of the cone, so that the parameters d and Δ are assumed to be non-essential (compare with G.I. Taylor's idealization of the explosion problem in the Introduction).

In the *LFT* class we obtain for the dimensions of the quantities involved

$$[D] = L, \qquad [P] = F, \qquad [K] = \frac{F}{L^{3/2}}, \qquad [v] = 1.$$

(The dimension of the cohesion modulus K is obtained from the relation giving the stress intensity factor N, $N \sim \sigma \sqrt{s}$, and the dimension of the stress, which is the same as the dimension of the pressure, $[\sigma] = FL^{-2}$.)

The governing parameters P and K have independent dimensions, and the standard procedures of dimensional analysis immediately yield

$$\Pi = \frac{D}{P^{2/3}K^{-2/3}} = \Phi(v), \qquad \text{i.e. } D = \left(\frac{P}{K}\right)^{2/3} \Phi(v).$$

Benbow's (1960) analysis of experimental data obtained for punches of various sizes under various loads confirmed this relation (Figure 1.8).

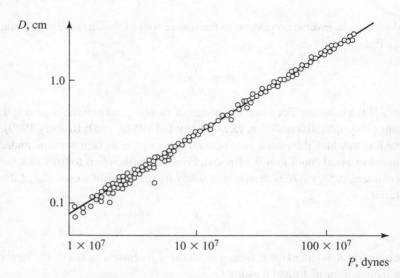

Figure 1.8. The experimental data on the propagation of a cone crack in a block of fused silica confirm the scaling law $D = \text{const}(P/K)^{2/3}$ (Benbow 1960).

The five examples just discussed demonstrate that the seemingly trivial concepts of dimensional analysis are capable of producing results with a great deal of content, especially when the difference between the the total number of governing parameters and the number of governing parameters with independent dimensions is not large. Thus, correctly choosing the set of governing parameters becomes the most important factor: it is important not only to take all essential variables into account but also to exclude superfluous ones! The set of governing parameters may be determined relatively easily if a mathematical formulation of the problem is available.[9] This must include the governing variables and constant parameters of the problem, which appear in the equations, boundary conditions, initial conditions and so forth and which determine the unique solution to the problem. Correctly choosing the set of governing parameters for problems that do not have an explicit mathematical formulation depends primarily on the intuition of the researcher. In such problems, success in applying dimensional analysis involves a correct understanding of which governing parameters are essential and which may be neglected. Remember that each governing parameter that can be neglected reduces the amount of work involved in investigating the problem by roughly an order of magnitude!

[9] As we shall see later, there are many subtle points even here.

We note in concluding this section that our presentation of dimensional analysis is essentially different from those available in the literature, although it follows in its general ideas the excellent book by P.W. Bridgman (1931).

1.3 Physical similarity

1.3.1 Physically similar phenomena

Before a large, expensive object (for example, a ship or aircraft) is constructed, experimentation on models – *modelling* – is used to determine the best properties under future operating conditions. Many different kinds of measurement are carried out on models: for example, the lift and drag of an aircraft model as air flows past it can be measured in a wind tunnel, as can the aerodynamic loading that causes a model of a television tower to collapse. Clearly, one must know how to scale the results of the experiment carried out on the model up to the full-scale object being modelled. If one does not know how to do this, modelling is a useless pursuit. The concept of *physically similar phenomena* is central to correct modelling.

The concept of physical similarity is a natural generalization of the concept of similarity in geometry. For example, two triangles are similar if they differ only in the numerical values of the dimensional parameters, i.e. the lengths of the sides, while the dimensionless parameters, the angles at the vertices, are identical for the two triangles. Analogously, *physical phenomena are called similar if they differ only in respect of the numerical values of the dimensional governing parameters, the values of the corresponding dimensionless parameters* $\Pi_1, \ldots, \Pi_m$ *being identical.* In accordance with this definition, the quantities $\Pi_1, \ldots, \Pi_m$ are called *similarity parameters*.

We shall now imagine that we propose to model a certain phenomenon; we shall call this phenomenon the *prototype*. We require that the *model* that we use to determine the desired properties of the prototype be a phenomenon *physically similar* to the prototype. We have for both phenomena the relationship (1.19) between the parameter a to be determined and the governing parameters $a_1, \ldots, a_k, b_1, \ldots, b_m$:

$$a = f(a_1, \ldots, a_k, b_1, \ldots, b_m).$$

The function f is the same for both phenomena because we require them to be similar, though the numerical values of the governing parameters $a_1, \ldots, a_k$, $b_1, \ldots, b_m$ and the parameter a to be determined may differ. Thus, relationship (1.19) for the prototype takes the form

$$a^{\mathrm{P}} = f\left(a_1^{\mathrm{P}}, \ldots, a_k^{\mathrm{P}}, b_1^{\mathrm{P}}, \ldots, b_m^{\mathrm{P}}\right). \tag{1.40}$$

The superscript P will hereinafter be used to refer to the properties of the prototype. Relation (1.19) for the model is similar in form, but the numerical values of the governing and determined parameters are different:

$$a^M = f\left(a_1^M, \ldots, a_k^M, b_1^M, \ldots, b_m^M\right). \tag{1.41}$$

The superscript M will hereinafter be used to refer to the properties of the model. Via dimensional analysis, (1.24), we obtain

$$\Pi^P = \Phi\left(\Pi_1^P, \ldots, \Pi_m^P\right), \qquad \Pi^M = \Phi\left(\Pi_1^M, \ldots, \Pi_m^M\right), \tag{1.42}$$

where the function Φ is the same in both cases, since it can be expressed in terms of the function f in the same way in each case; Π^P, Π^M and the Π_i^P and Π_i^M are dimensionless parameters.

1.3.2 The rule for scaling the results for a physically similar model up to the prototype

Consider two similar triangles, one the prototype and one a model. For both of them the dimensionless parameters – the angles at the vertices, $\Pi_1 = \alpha$ and $\Pi_2 = \beta$, – are identical whereas the only dimensional governing parameter, say the side opposite to angle α, of length ℓ, takes different values, ℓ^P and ℓ^M. The areas of the prototype and model triangles are expressed as

$$S^P = (\ell^P)^2 \Phi(\alpha, \beta), \qquad S^M = (\ell^M)^2 \Phi(\alpha, \beta),$$

so that

$$S^P = S^M \left(\frac{\ell^P}{\ell^M}\right)^2.$$

A similar *scaling rule* can be obtained in the general case of physically similar phenomena.

Since the model and prototype are similar, the following conditions on the dimensionless governing parameters must be satisfied, according to the definition of physically similar phenomena given above:

$$\Pi_1^M = \Pi_1^P, \qquad \ldots, \qquad \Pi_m^M = \Pi_m^P. \tag{1.43}$$

Conditions (1.43) are called the *similarity criteria*.

Hence, as stated above,

$$\Phi\left(\Pi_1^M, \ldots, \Pi_m^M\right) = \Phi\left(\Pi_1^P, \ldots, \Pi_m^P\right);$$

also, in accordance with (1.42), the dimensionless parameters to be determined for the model and for the prototype are equal:

$$\Pi^P = \Pi^M. \tag{1.44}$$

Returning to the dimensional parameters $a, a_1, \ldots, a_k$ using (1.22), we find that

$$a^P = a^M \left(\frac{a_1^P}{a_1^M}\right)^p \cdots \left(\frac{a_k^P}{a_k^M}\right)^r, \tag{1.45}$$

which is a *simple rule for scaling the results of measurements on a similar model up to the prototype.* It was precisely in order to be able to use this relationship that it was necessary to require that the model be similar to the prototype.

1.3.3 Choosing the governing parameters of the model

The model parameters $a_1^M, \ldots, a_k^M$ may be selected arbitrarily, keeping in mind maximum simplicity and convenience in modelling. Then the conditions for similarity between the model and the prototype – equality of the similarity parameters $\Pi_1, \ldots, \Pi_m$ for model and prototype, (1.43) – show how the remaining governing model parameters $b_1^M, \ldots, b_m^M$ must be chosen in order to ensure similarity between model and prototype. These conditions are as follows.

$$\Pi_1^M = \Pi_1^P \quad \Rightarrow \quad b_1^M = b_1^P \left(\frac{a_1^M}{a_1^P}\right)^{p_1} \cdots \left(\frac{a_k^M}{a_k^P}\right)^{r_1};$$

$$\vdots \qquad\qquad \vdots \tag{1.46}$$

$$\Pi_m^M = \Pi_m^P \quad \Rightarrow \quad b_m^M = b_m^P \left(\frac{a_1^M}{a_1^P}\right)^{p_m} \cdots \left(\frac{a_k^M}{a_k^P}\right)^{r_m}.$$

The simple definitions and statements presented above describe the entire content of the theory of similarity: we emphasize that there is nothing more to this theory. The examples given below will demonstrate how to use the theory. Along the way, the reader will become familiar with the most important classical similarity parameters.

1.3.4 Problems

Problem 1. *Derive the rules for modelling the steady motion of a body in a fluid that fills a very large vessel.*

The velocity of the body is assumed to be small in comparison with the velocity of sound in the fluid. Therefore the compressibility of the fluid may be neglected, and its density is assumed to be constant.

Solution

(a) *Geometric and kinematic similarity conditions.* The model's body shape must be geometrically similar to the prototype's body shape, and the direction of the velocity vector with respect to the principal axes of the body must be identical in the model's motion and in the prototype's motion.[10]

(b) *Dynamic similarity condition.* The dimensional governing parameters of the motion are the characteristic length scale of the body, its maximum cross-sectional diameter D, for example, the magnitude U of the body's velocity, the density ρ of the fluid and its viscosity μ. The dimensions of the governing parameters in the *LMT* class are as follows:

$$[D] = L, \qquad [U] = \frac{L}{T}, \qquad [\rho] = L^{-3}M, \qquad [\mu] = L^{-1}MT^{-1}. \quad (1.47)$$

Clearly the number of parameters with independent dimensions $k = 3$, so that there is only one dynamical similarity parameter in addition to the geometric similarity parameters (which express the similarity of the model's and prototype's body shapes) and the kinematic similarity parameters (which express the identical orientation of the velocity with respect to the principal axes of the model and of the prototype); this parameter can be written in the following form (cf. subsection 1.2.3):

$$\Pi_1 = \frac{\rho U D}{\mu} = Re. \quad (1.48)$$

As proposed by A. Sommerfeld, this parameter is called the *Reynolds number* or *Reynolds parameter*.

The dimensionless drag force Π acting on the body can be defined naturally in the following way:

$$\Pi = \mathcal{F} \Big/ \left(\frac{1}{2}\rho U^2 S\right).$$

Here $\mathcal{F}$ is the drag force acting on the body, $S \sim D^2$ is the cross-sectional area of the moving body and the factor $\frac{1}{2}$ is introduced by convention.

The function $\Pi(Re)$ for the flow past a sphere is shown in Figure 1.9; to good accuracy, the data from a large number of experiments lie on a single curve.

---

[10] This follows from the identity of the corresponding geometric and kinematic similarity parameters for the model and prototype.

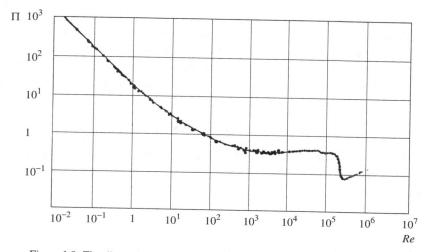

Figure 1.9. The dimensionless drag force on a sphere as a function of Reynolds number. The data from the various experiments shown here turn out to lie on a single curve, which indicates that the Reynolds number is the only parameter that governs the global structure of the flow. The complicated nature of the curve indicates that the flow regime changes with Reynolds number.

This curve appears to be very complicated: a large region in which $\Pi(Re)$ varies smoothly gives way to sudden decreases and finally an increase, and there are two regions where Π is almost independent of Re. This all indicates that the actual flow regimes vary with Reynolds number: Re is the only parameter that governs the structure of the flow as a whole as the sphere is moving in the fluid.

The motion of a model is usually implemented in the same fluid as that in which the prototype moves. The similarity condition that parameter (1.48) be the same for the model and prototype motions indicates that, in this case, the product UD must be identical for model and prototype: from this, we see that the ratio of the model and prototype velocities must be the inverse of the ratio of their linear dimensions. From this it follows that the drag forces must be identical for the model and the prototype, so that the scaling coefficient for the drag force is equal to unity in this case.

Problem 2. *Derive the rules for modelling the steady motion of a streamlined surface ship at high speeds.*

We assume that the main contribution to the drag of a streamlined surface ship in rapid motion comes from surface waves created by the ship. The contribution from viscous drag for such a situation is assumed, in a rough first approximation, to be small in comparison with the wave drag, so that it can be neglected.

Solution. The governing parameters in the case at hand will be as follows: a characteristic length for the ship, ℓ, the gravitational acceleration g, the density of the fluid ρ and the speed of the ship U. The parameter g is essential since the gravitational force is one of the factors that controls the waves created by the ship. The governing parameters have the following dimensions in the *LMT* class:

$$[\ell] = L, \qquad [g] = LT^{-2}, \qquad [\rho] = L^{-3}M, \qquad [U] = LT^{-1}, \qquad (1.49)$$

so that $k = 3, m = 1$, and the only dynamical similarity parameter (in addition to the geometric and kinematic similarity parameters, see the previous problem) is of the form

$$\Pi_1 = \frac{U}{\ell^{1/2} g^{1/2}}. \qquad (1.50)$$

This parameter is called the *Froude number* or *Froude parameter* (the conventional symbol is *Fr*) after the English shipbuilder William Froude.

Furthermore, the dimension of the drag force $\mathcal{F}$ in the same class, *LMT*, is $[\mathcal{F}] = LMT^{-2}$, so that $[\mathcal{F}] = [\rho][g][\ell]^3$. Thus, since the parameter g can be varied only with a great deal of effort, by means of subtle tricks not normally used, the law for scaling the drag force from the model up to the prototype in the same fluid is of the form

$$\mathcal{F}^{\mathrm{P}} = \mathcal{F}^{\mathrm{M}} \left(\frac{\ell^{\mathrm{P}}}{\ell^{\mathrm{M}}} \right)^3, \qquad (1.51)$$

so that the drag force is proportional to the cube of the modelling scale. Relation (1.50) indicates that, in order to ensure dynamical similarity, the ratio of the model velocity to the prototype velocity must be proportional to the square root of the modelling scale:

$$U^{\mathrm{M}} = U^{\mathrm{P}} \left(\frac{\ell^{\mathrm{M}}}{\ell^{\mathrm{P}}} \right)^{1/2}. \qquad (1.52)$$

If one does not neglect the role of viscosity, a second dynamical parameter appears – the Reynolds number *Re*, which in the present notation is equal to $\rho U \ell / \mu$. Modelling in which both similarity parameters – the Froude number and Reynolds number – are taken into account turns out to be impossible in a single fluid. Indeed, to do this, the products $U\ell$ (see the previous problem) and U^2/ℓ would have to be identical for both model and prototype; this is only possible when modelling to full scale, which makes no sense. This is precisely why, for illustration, we have restricted ourselves to the case where the viscous drag is small compared with the wave drag. Thus, in ship-building practice

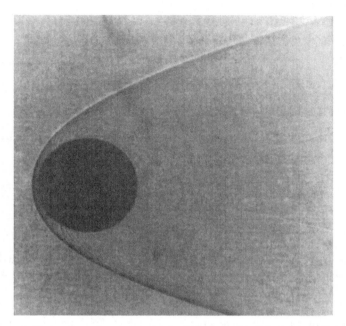

Figure 1.10. A nylon sphere moves in air at Mach number 7.6. A detached shock wave is visible ahead of the sphere (van Dyke 1982).

the viscous-drag contribution is modelled separately from the wave drag using specially developed techniques.

Problem 3. *Derive the rules for modelling the steady motion at high velocity of a body in a gas*

Consider the simplest case: the motion of a sphere, an example of which is shown in Figure 1.10. The drag acting on the sphere is obtained on the small model and then scaled up to the prototype.

Solution. The drag $\mathcal{F}$ depends on the following governing parameters: the gas density ρ, the velocity of the sphere U, its diameter D, and the velocity of sound c:

$$\mathcal{F} = f(\rho, U, D, c). \qquad (1.53)$$

The dimensions of the quantities involved in the *LMT* class are

$$[\mathcal{F}] = \frac{ML}{T^2}, \qquad [\rho] = \frac{M}{L^3}, \qquad [U] = \frac{L}{T}, \qquad [D] = L, \qquad [c] = \frac{L}{T}.$$
$$(1.54)$$

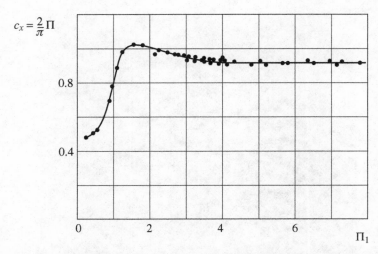

Figure 1.11. The dimensionless drag on a sphere, Π (times $2/\pi$), as a function of the dimensionless governing parameter $\Pi_1 = U/c$, the Mach number (Chernyi 1961). The quantity Π approaches a constant for large values of Π_1.

Three first governing parameters have independent dimensions. The standard procedure of dimensional analysis gives

$$\Pi = \frac{\mathcal{F}}{\rho U^2 D^2} = \Phi(\mathcal{M}). \tag{1.55}$$

Here

$$\mathcal{M} = \frac{U}{c} = \frac{1}{\Pi_1} \tag{1.56}$$

is the Mach number, named in honour of the Austrian scientist who performed pioneering experiments with shock waves in a gas. The function $\Phi(\mathcal{M})$, for the flow around a sphere, was obtained from experiments in a wind tunnel (Figure 1.11). It is interesting to note that for large $\mathcal{M}$, in fact for $\mathcal{M} \geq 4$, the function $\Phi(\mathcal{M})$ is close to a constant so that then the drag $\mathcal{F}$ is determined by the relation

$$\mathcal{F} = \text{const } \rho U^2 D^2. \tag{1.57}$$

In fact very often, if a dimensionless parameter is large or small, the dependence upon it is neglected too soon and the corresponding dimensional parameter, in this case the sound velocity c, is dropped. In the special case which we have considered this procedure is correct but as a rule, and as we will see further, it is not.

Problem 4. *Derive the rules for modelling thermal convection in a horizontal fluid layer.*

We assume that the layer is bounded by smooth rigid isothermal walls: at the upper wall a temperature T_0 is maintained and at the lower wall a higher temperature, $T_0 + \delta T$.

Solution. The phenomenon of convection in the gravity field is due to the fact that the density of a fluid usually decreases as it is heated; if this decrease is large enough, the less dense fluid floats from bottom to top. We shall neglect the variations with temperature in the viscosity, specific heat capacity and thermal conductivity of the fluid because we intend to model the process of convection in a basic way, not in detail. For small changes in temperature, the temperature dependence of the fluid density can be assumed to be linear:

$$\rho = \rho_0[1 + \alpha(T_0 - T)],$$

where ρ_0 is the fluid density at temperature T_0 and α is the coefficient of volume expansion of the fluid. The variation in the density of the fluid as it is heated is small, so that we need only take the density variation into account where it is combined with the action of the gravitational force. This approximation was suggested by the French scientist J. Boussinesq, and carries his name. The *Boussinesq approximation* is related to the assumption that all the accelerations in convective flow are small compared with the gravitational acceleration. This is not so in strongly developed convection; here the Boussinesq approximation is no longer valid. If we adopt the Boussinesq approximation then the coefficient of volume expansion of the fluid, α, and the gravitational acceleration g do not enter into consideration separately but only as a product. The product αg is called the *buoyancy parameter.*

The governing parameters for the phenomenon of thermal convection in a layer are as follows. The properties of the phenomenon must depend on the buoyancy parameter αg, on the thickness of the layer H, on the dynamical properties of the fluid, that is, its viscosity μ and density ρ_0 at temperature T_0, on its specific heat capacity c and thermal conductivity λ and on the excess temperature of the lower layer δT.

In principle, the contribution of viscous energy dissipation to the thermal balance of the fluid should also be taken into account. To do this, one additional parameter must be included, the mechanical equivalent of heat J (cf. the discussion at the end of subsection 1.1.3).

The dimensions of the governing parameters can be obtained in the following way. The specific heat capacity c is, by definition, the quantity of heat necessary to increase the temperature of a unit mass of the fluid by one temperature unit. Thus, the dimension of heat capacity is

$$[c] = \frac{Q}{M\Theta},\tag{1.58}$$

where Q stands for the independent dimension of the quantity of heat and Θ stands for the independent dimension of temperature. The thermal conductivity of the fluid, λ, is, by the fundamental law of heat conduction (the Fourier law), the coefficient of proportionality in the expression for the heat flux through a horizontal layer of quiescent fluid as a function of the temperature drop δT and the thickness of the layer H:

$$q = -\lambda \frac{\delta T}{H}.$$

Now the heat flux is, by definition, the amount of heat that passes through unit area of the plane layer boundary per unit time, so that $[q] = QL^{-2}T^{-1}$. From this result and the preceding equation, we find that

$$[\lambda] = \frac{Q}{LT\Theta}.\tag{1.59}$$

The dimension of the mechanical equivalent of heat is obviously equal to the dimension of mechanical energy divided by the independent dimension of thermal energy:

$$[J] = \frac{ML^2}{T^2 Q}.\tag{1.60}$$

So, the dimensions of the governing parameters in the $LMT\Theta Q$ class are as follows:

$$[\alpha g] = \frac{L}{\Theta T^2}, \quad [H] = L, \quad [\mu] = \frac{M}{LT}, \quad [\rho] = \frac{M}{L^3},$$

$$[c] = \frac{Q}{M\Theta}, \quad [\delta T] = \Theta, \quad [\lambda] = \frac{Q}{LT\Theta}, \quad [J] = \frac{ML^2}{T^2 Q}.\tag{1.61}$$

Clearly $n = 8$, $m = 3$ and $k = 5$. Applying dimensional analysis, we obtain the following three similarity parameters:

$$\Pi_1 = \frac{\delta T}{(\alpha g)^{-1} H^{-3} \mu^2 \rho^{-2}}, \quad \Pi_2 = \frac{\lambda}{\mu c}, \quad \Pi_3 = \frac{Jc}{\alpha g H}.\tag{1.62}$$

In what follows, we shall discuss convective motion in thin layers, for which the parameter Π_3 is large ($\Pi_3 \gg 1$), so that the effect of this parameter on the

Figure 1.12. Uniform heating from below a fluid in a vessel shaped like a rectangular parallelpiped with sides in the ratios $10:4:1$ produces flow with rotating rolls parallel to one of the sides. The rolls with dark and light centers rotate in opposite directions. From van Dyke (1982).

similarity conditions may be neglected.[11] It is useful to estimate the value of the characteristic length $\Lambda = Jc/(\alpha g)$ in order to get an idea of the extent to which this condition is restrictive. We have $J = 4.2 \times 10^7$ erg/cal, $c = 1$ cal/g °C and $\alpha = 2 \times 10^{-4}/°K$ for water, and $g = 10^3$ cm/s^2, from which we find that $\Lambda \simeq 2 \times 10^8$ cm $= 2000$ km. Thus, when modelling convection in water, even a layer one kilometer deep can be assumed to be thin, and so Π_3 can be neglected. However, when modelling convection in the Earth's mantle the parameter Π_3 is of order unity and cannot be neglected.

We should also note one significant fact that follows from the relations (1.62) for the similarity parameters. If the contribution to the thermal balance from energy dissipation in the convective motion is neglected, it turns out that the quantities λ and c enter into the discussion as a ratio rather than separately.

The parameter Π_1 is called the *Grasshof number*, and the following combinations of the parameters in (1.62) are frequently used in the literature:

$$\frac{\Pi_1}{\Pi_2} = \frac{\alpha g\, \delta T H^3}{\mu \rho^{-2} \lambda c^{-1}} = Ra, \qquad \frac{1}{\Pi_2} = \frac{\mu c}{\lambda} = Pr. \qquad (1.63)$$

The parameter Ra, the *Rayleigh number*, is named after the great English physicist who was the first to study the onset of convection in a horizontal layer theoretically. When a critical value of this parameter,

$$Ra = Ra_{\mathrm{cr}} \simeq 657, \qquad (1.64)$$

is reached (Ra_{cr} does not depend on the second parameter,[12] the *Prandtl number Pr*) the state in which the fluid in a horizontal layer is at rest becomes unstable, and the so-called regime of buoyancy-driven convection rolls sets in. In this regime, the layer breaks up into fluid rolls that rotate in opposite directions (Figure 1.12). Until the Rayleigh number reaches this critical value, however,

[11] We repeat that the correctness of neglecting the effect of a certain parameter is always a strong assumption, no matter how large or small this parameter may be.

[12] The critical value given here is calculated under the assumption that the tangential stresses vanish at the boundaries of the layer.

the equilibrium state for a quiescent fluid layer is stable. Later changes in the convection regime in the horizontal layer are associated with the passage of the Rayleigh number through other critical values.

The similarity parameters (1.62) indicate that if modelling is carried out in the same fluid and same gravitational field as pertains for the prototype motion then the following condition on the model's temperature difference δT^M must be satisfied:

$$\delta T^M = \delta T^P \left(\frac{H^P}{H^M} \right)^3 . \tag{1.65}$$

This condition ensures that the model's convective motion is physically similar to that of the prototype. Furthermore, as may easily be shown, the dimensionless parameter for the heat flux q is of the form

$$\Pi = \frac{q}{(\alpha g)^{-1} H^{-4} \mu^3 \rho^{-2} c}. \tag{1.66}$$

Thus, the rule for scaling the heat flux when modelling in a layer of the same fluid as the prototype motion takes the following form:

$$q^P = q^M \left(\frac{H^M}{H^P} \right)^4 , \tag{1.67}$$

so that the ratio of the heat fluxes in the prototype and in the model must be inversely proportional to the fourth power of the modelling scale.

As was mentioned above, the influence of the similarity parameter Π_3 becomes appreciable for thick layers. Since this parameter is given by the ratio of the characteristic length scale of the fluid, $\Lambda = Jc/(\alpha g)$, and the layer thickness-, it is strictly speaking impossible to model the phenomenon in a layer of the same fluid under identical external conditions (compare this result with the second problem).

The present example shows that one must be careful when determining the similarity parameters. For example, if we assume that the dimensions of mechanical energy and thermal energy are independent and therefore do not take the governing parameter for the mechanical equivalent of heat into consideration then we will not notice the restrictions on the thickness of the model layer. Meanwhile, the phenomena for thick and thin layers are substantially different; they are not physically similar, and it is in general not possible to scale the heat fluxes using the simple relation in (1.67).

Furthermore, it is obvious that if the thermal and mechanical energy were measured in the same units, i.e., if we were to pass from the $LMT\Theta Q$ class to the $LMT\Theta$ class, the conclusions reached above would not be affected in any respect. Indeed, the difference between the total number of governing parameters and the number of governing parameters with independent dimensions would remain constant, even though the mechanical equivalent of heat had been removed from consideration. A negligible contribution of viscous dissipation to the thermal balance would then be interpreted to mean that the phenomenon is not governed by the heat capacity c and thermal conductivity λ separately. The governing parameter would be their ratio, which appears in the so-called thermometric conductivity $\kappa = \lambda/(\rho c)$. This would lead to a decrease in the number of governing parameters and the disappearance of the similarity parameter Π_3.

Problem 5. *Derive the rules for modelling the steady motion of row boats. Compare the velocities of boats accommodating a varying number n of oarsmen, one, two, four and eight* (McMahon 1971).

Solution. It seems natural to make the following assumptions. (a) There is a geometric similarity between the boats. (b) The volume of a loaded boat per oarsman G is a constant, characteristic for boats of all classes. This follows from assuming that the bulk weight of the boat per oarsman, including the oarsman's own weight, is constant; we shall consider the oarsmen to be indistinguishable in weight. (c) The power per oarsman A is a constant, characteristic for all classes, so the oarsmen are considered as indistinguishable also in power.

The principal force $\mathcal{F}$ that hinders the motion of the boat through the water is, unlike in the previous exercise, skin friction drag. Indeed, full-scale rowing-tank tests have shown that the resistance due to leeway and wave-making constitute together only a tiny part of the total drag. In the range of Reynolds numbers characteristic for racing, the 'drag coefficient'

$$\lambda = \frac{\mathcal{F}}{\rho v^2 \ell^2}$$

can be considered as a constant. Here ρ is the water density, v is the velocity of motion, assumed to be steady, and ℓ is the characteristic length scale of the wetted surface. Therefore the bulk power supporting the motion is

$$P = \mathcal{F}v = \lambda \rho v^3 \ell^2. \tag{1.68}$$

Obviously the bulk power is proportional to the number n of oarsmen: $P = An$.

Thus, the velocity of the motion v is a function of the governing parameters n, A, G and ρ. The dimensions of the parameters in the class $RVLN$ (R is the dimension of density, V is the dimension of velocity, L is the dimension of

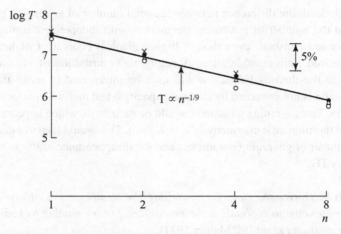

Figure 1.13. The $-1/9$ power-law dependence of the rowing time T on the number of oarsmen n (solid line). This may be compared with racing times over 2000 m, all at calm or near calm conditions: $\triangle$, 1964 Olympics, Tokyo; $\bullet$, 1968 Olympics, Mexico City; $\times$, 1970 World Rowing Championships, Ontario; $\circ$, 1970 Lucerne International Championships. After McMahon (1971).

length and N is the dimension of the number n of oarsmen;[13] these four can be considered as independent dimensions) are

$$[v] = V, \qquad [G] = \frac{L^3}{N}, \qquad [A] = RV^3L^2N^{-1}, \qquad [\rho] = R, \qquad [n] = N. \tag{1.69}$$

So, again referring back to (1.19), $m = 0$ and $k = 4$. Using dimensional analysis we obtain

$$\Pi = \text{const}, \qquad \Pi = \frac{v}{A^{1/3}\rho^{-1/3}G^{-2/9}n^{1/9}} \tag{1.70}$$

and the final result for the velocity of the boat is

$$v = \text{const} \, \frac{A^{1/3}}{\rho^{1/3}G^{2/9}} \, n^{1/9}. \tag{1.71}$$

According to (1.71) the time for a fixed distance, say 2000 m, should be inversely proportional to $n^{1/9}$. The validity of this result is well illustrated by Figure 1.13.

---

[13] The units of the number of oarsmen in a boat are squad/N (or batallion/N etc.). Military units are used and it is assumed that there are a fixed number of privates in each unit.

The examples presented above[14] show that dimensional considerations play a decisive role in establishing rules for modelling and criteria for similarity. The crucial step in modelling, as in any application of dimensional analysis to cases where an exact mathematical formulation of the problem is missing, lies in the proper choice of a system of governing parameters. Often the procedure is as follows. An investigator takes as governing parameters all quantities that, in his or her opinion, could possibly have an influence on the phenomenon, no matter how hypothetical. As governing parameters with independent dimensions the investigator takes those governing parameters that are definitely known to be essential, and, with respect to the remaining ones, he or she looks at the numerical values of the corresponding similarity parameters Π_i. If these values are very small or very large, the corresponding dimensional parameter b_i is somewhat recklessly considered inessential and is discarded.

In many cases one can actually proceed in this way but it is very important to note that in general this is *not so*, and one must be very careful about arguments such as the above. One should see in them not a proof of the possibility of disregarding one parameter or another but a strong conjecture. This last assertion is obvious: it is not necessarily true that a function $\Pi = \Phi(\Pi_1, \ldots, \Pi_i, \ldots, \Pi_m)$ tends to a definite and moreover finite non-zero limit for small or large values of the argument Π_i. Only the existence of such a limit (and in fact even of a sufficiently rapid convergence to it) can justify neglecting a governing parameter when the corresponding similarity parameter is very large or very small. Subsequent discussion will show us that crudeness of analysis here can lead to serious mistakes.

---

[14] Very instructive also are similarity considerations for atmospheres of planets (Golitsyn 1973).

Chapter 2
Self-similarity and intermediate asymptotics

The statement that a certain phenomenon is *steady*, i.e. time independent, is obviously very significant: there is then no need to trace its evolution in time. Of similar significance is the statement that the phenomenon is *self-similar*. This means that the *spatial distributions of the characteristics of the phenomenon* (such as flow velocity, stress, electric current, etc.), $\mathbf{U}(\mathbf{r}, t)$, *vary with time while remaining geometrically similar*. In other words, there exist time-dependent scales $U_0(t)$ and $r_0(t)$ such that measured in these scales the phenomenon becomes time independent:

$$\mathbf{U}(\mathbf{r}, t) = U_0(t)\mathbf{f}\left(\frac{\mathbf{r}}{r_0(t)}\right).$$

We considered a remarkable self-similar phenomenon in the Introduction: the propagation of very intense blast waves. We will consider here, in detail, groundwater flow after very intense flooding. Our goal is to demonstrate, using this transparent and mathematically very simple example, the application of dimensional analysis for constructing self-similar solutions to the partial differential equations of mathematical models. In particular, we want to demonstrate the principal difficulties which can arise in such an application.

2.1 Gently sloping groundwater flow. A mathematical model

Consider (Figure 2.1) a stratum consisting of porous rock, for example, sandstone, on top of an underlying horizontal impermeable bed and containing a dome of groundwater or some other liquid, for example, liquid waste. The rock is assumed to be a porous medium which is permeable for fluid flow. Therefore, under the influence of gravity the groundwater dome will spread out along the impermeable bed.

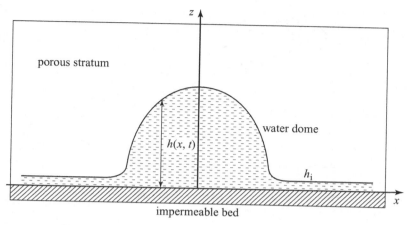

Figure 2.1. A groundwater dome is formed in a porous stratum.

We will consider the one-dimensional case, when all flow characteristics depend on the horizontal space coordinate x and the time t only.

The gently sloping fluid motion in the porous medium is slow, so that the water pressure within the dome may be assumed to obey the hydrostatic law $p = \rho g(h - z)$, where h is the groundwater level, ρ is the fluid density, g is the gravitational acceleration and z is the vertical coordinate reckoned from the bed. (We neglect the pressure of the gas in the porous medium above the dome.) Thus the total 'head', i.e. the convenient quantity $H = p + \rho g z$ (here and hereafter I depart from the traditional terminology used in hydraulics, where the head is called the quantity $H/\rho z$), remains constant and equal to $\rho g h$ throughout every vertical column of height h in the groundwater dome.

The fundamental law governing slow motions in a porous medium is the Darcy law, named after the French engineer who discovered it in the middle of the nineteenth century while investigating the public fountains in Dijon. Darcy's law claims that (for details see the classic books Polubarinova-Kochina 1962; Bear 1972) the flux of fluid (the flow rate per unit time *and per unit total cross-sectional area*[1]) is proportional to the gradient of the total head and inversely proportional to the fluid viscosity μ. In the case under consideration, of gently sloping flow, the groundwater head H is constant throughout any vertical in the dome: $H = H(x, t)$ and according to Darcy's law the total flux q through an area of unit width is equal to

$$q = -\frac{k}{\mu} h \partial_x H = -\frac{k}{\mu \rho g} H \partial_x H. \qquad (2.1)$$

---

[1] It is essential that the total cross-sectional area is considered, not just the cross-sectional area of the pores open for fluid motion.

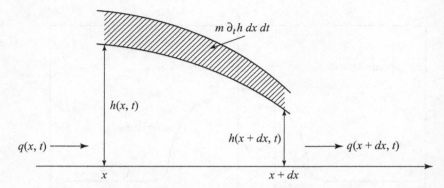

Figure 2.2. Diagram for deriving the equation describing gently sloping ground-water motion.

Here k is the *permeability coefficient*, a constant property of a porous medium. It has the dimension of area (squared length) and its magnitude usually lies in the range $10^{-9} - 10^{-8}$ cm^2. Further, we will use another property of the porous medium, its *porosity m*, the fractional volume of the medium occupied by pores; usually m is of order 10^{-1}. It can be shown (Polubarinova-Kochina 1962; Bear 1972) that the fractional area of the pore cross-sections also can be taken as equal to m. In calculating the flux we have divided the flow rate by the total area, which is partly occupied by pore cross-sections. Therefore the average *fluid velocity* in the pores is equal to

$$u = -\frac{1}{m}\frac{k}{\mu}\partial_x H. \tag{2.2}$$

To derive the basic equation for the head H, or that for the groundwater level h, we consider a section of the stratum between adjacent cross-sections at x and $x + dx$ (Figure 2.2). The change in the quantity of water in this section during the time interval between t and $t + dt$, indicated by the shaded area, is equal to $m \, \partial_t \, h \, dx \, dt$. However, it is also equal to the difference $-\partial_x \, q \, dx \, dt$ between the fluxes over the cross-sections at x and $x + dx$ because the inflow or outflow within the volume of the section between them is assumed to vanish, i.e. for the present we assume no absorption. Therefore we obtain, using expression (2.1) for the total flux, the equation

$$m\partial_t h = \frac{k}{\mu\rho g} \, \partial_x(H\partial_x H) \quad \text{i.e.} \quad \partial_t H = \kappa \partial_{xx}^2 H^2 \tag{2.3}$$

where

$$\kappa = \frac{k}{2m\mu}. \tag{2.4}$$

Equation (2.3) is named the Boussinesq equation after the French scientist J. Boussinesq, who derived it in 1904. Equation (2.3) should be completed by an initial condition

$$H(x, 0) = H_0(x),\qquad(2.5)$$

where $H_0(x)$ is a non-negative function, and by appropriate boundary conditions. Equation (2.3) and the initial and boundary conditions form the general mathematical model of the phenomenon under consideration.

2.2 Very intense concentrated flooding: the self-similar solution

Assume that at the moment $t = 0$ a groundwater dome is formed very quickly, so we can consider it as formed *instantaneously*. The dome is formed between sections $x = -\ell$ and $x = \ell$, and it is *concentrated*, so that its horizontal extent 2ℓ is assumed to be small in comparison with the horizontal extension of the stratum. The initial weight of the dome is equal to

$$\int_{-\ell}^{\ell} H(x, 0)\, dx = I.\qquad(2.6)$$

The flooding is *very intense*, so that the mean initial water head in the dome $I/(2\ell)$ is assumed to be large in comparison with the uniform initial water head H_i in the stratum outside the dome. Without loss of generality it is possible to represent the initial groundwater-head distribution in the following form, useful for subsequent dimensional analysis:

$$H_0(x) = \frac{I}{\ell} f_0\left(\frac{x}{\ell}\right)\qquad(2.7)$$

where f_0 is a dimensionless function of its dimensionless argument. Thus the groundwater head at a subsequent time t depends on the arguments t, I, κ, H_i, ℓ and x, so that

$$H = f(t, I, \kappa, H_i, \ell, x).\qquad(2.8)$$

It is easy to show that the dimensions of the involved quantities are as follows:

$$[H] = [H_i] = \frac{M}{LT^2}, \qquad [t] = T, \qquad [I] = \frac{M}{T^2},$$

$$[\kappa] = \frac{L^3 T}{M}, \qquad [\ell] = [x] = L.\qquad(2.9)$$

Three of the governing, parameters, t, I and κ, have independent dimensions. The dimensions of the remaining parameters can be expressed in terms of the dimensions of the three parameters with independent dimensions:

$$[H] = [H_i] = [t]^{-1/3}[I]^{2/3}[\kappa]^{-1/3}, \qquad [x] = [\ell] = \left[(I\kappa t)^{1/3}\right]. \qquad (2.10)$$

The standard procedure of dimensional analysis, described in Chapter 1, gives the following results:

$$\Pi = \frac{H}{I^{2/3}\kappa^{-1/3}t^{-1/3}} = \Phi(\Pi_1, \Pi_2, \Pi_3),$$

$$\Pi_1 = \frac{x}{(I\kappa t)^{1/3}}, \qquad \Pi_2 = \frac{\ell}{(I\kappa t)^{1/3}}, \qquad \Pi_3 = \frac{H_i}{I^{2/3}\kappa^{-1/3}t^{-1/3}} \qquad (2.11)$$

so that

$$H = \frac{I^{2/3}}{(\kappa t)^{1/3}} \, \Phi\left(\frac{x}{(I\kappa t)^{1/3}}, \frac{\ell}{(I\kappa t)^{1/3}}, \frac{H_i}{I^{2/3}(\kappa t)^{-1/3}}\right). \qquad (2.12)$$

This result seems disappointing: instead of the two independent variables, x and t, in the original problem formulation we have obtained, after using dimensional analysis, three – a complication rather than a simplification of the problem! However, following exactly the example of G.I. Taylor (see the Introduction), we can replace the problem by an idealized one: *assume that the flooding is concentrated in the section $x = 0$, so that $\ell = 0$, and assume also that the initial water head in the stratum is negligible, so that $H_i = 0$* (the flooding is supposed to be very intense and concentrated). Then we obtain

$$H = \left(\frac{I^2}{\kappa t}\right)^{1/3} \Phi(\xi, 0, 0) = \left(\frac{I^2}{\kappa t}\right)^{1/3} F(\xi), \qquad \xi = \frac{x}{(I\kappa t)^{1/3}}, \qquad (2.13)$$

i.e. *the solution of the idealized problem has the property of self-similarity; see the equation at the start of the chapter.* Substituting (2.13) into the basic equation (2.3) we obtain an ordinary differential equation for the function $F(\xi)$:

$$\frac{d^2(F^2)}{d\xi^2} + \frac{\xi}{3}\frac{dF}{d\xi} + \frac{F}{3} = 0. \qquad (2.14)$$

This equation is in terms of total derivatives. We obtain easily by integration

$$\frac{dF^2}{d\xi} + \frac{\xi F}{3} = \text{const.} \qquad (2.15)$$

The constant on the right-hand side of (2.15) must be equal to zero since the solution must be finite and symmetric, so that $F'(0) = 0$. The next integration gives

$$F = \frac{1}{12}\left(\xi_f^2 - \xi^2\right), \qquad (2.16)$$

where for future convenience we denote the second constant of integration as $\xi_f^2/12$; the subscript refers to the front of the flow. Thus the solution to the ordinary differential equation (2.14) appears in the form

$$
F = \begin{cases} \dfrac{1}{12}(\xi_f^2 - \xi^2), & 0 \leq |\xi| \leq \xi_f \\ 0, & |\xi| \geq \xi_f \end{cases}
\tag{2.17}
$$

so that the solution (2.13) to be obtained can be represented in the form

$$
H = \begin{cases} \dfrac{1}{12}\left(\dfrac{I^2}{\kappa t}\right)^{1/3}\left(\xi_f^2 - \dfrac{x^2}{(I\kappa t)^{2/3}}\right), & 0 \leq |x| \leq x_f = \xi_f(I\kappa t)^{1/3} \\ 0 & |x| > x_f = \xi_f(I\kappa t)^{1/3} \end{cases}
\tag{2.18}
$$

Two points remain to be clarified: the value of the constant ξ_f has not yet been determined and we need to explain the finite discontinuity of the derivative $\partial_x H$ at the water fronts $x = \pm x_f$, where the head H vanishes. We start from the second point and note that for physical reasons both the head H and the total flux q given by (2.1) must be continuous. (Mathematically this means that no singularities such as $\delta(x - x_f)$ or $\delta'(x - x_f)$ on the right-hand side of (2.3) are allowed during the spreading of the mound.) Therefore, where $H \neq 0$ both H and $\partial_x H$ should be continuous. However, at the water fronts, $x = \pm x_f$, $H = 0$ and therefore the finite discontinuity of the derivative $\partial_x H$ that we obtained is allowed, because the flux $H\partial_x H$ at this point remains continuous. Indeed, at $|x| > x_f$ the head H is equal to zero (remember that an idealized problem where $H_i = 0$ is being considered) and therefore $H\partial_x H$ is zero at $|x| > x_f$. At the same time, according to (2.18) $H\partial_x H = 0$ at $x = x_f$. To determine the constant ξ_f we integrate equation (2.3) from $x = -x_f$ to $x = x_f$. We obtain

$$
\frac{d}{dt}\int_{-x_f}^{x_f} H(x,t)\,dx = \kappa\,\partial_x H^2\Big|_{-x_f}^{x_f} = 0,
\tag{2.19}
$$

so that the dome weight at time t,

$$
I(t) = \int_{-x_f}^{x_f} H(x,t)\,dx,
\tag{2.20}
$$

is constant in time. Therefore the dome weight is equal to its initial value I, and we obtain

$$
\int_{-x_f}^{x_f} \frac{1}{12}\left[\frac{I^2}{\kappa t}\right]^{1/3}\left[\xi_f^2 - \frac{x^2}{(I\kappa t)^{2/3}}\right]dx = I,
\tag{2.21}
$$

so that

$$\frac{\xi_f^2}{12} \int_{-\xi_f}^{\xi_f} \left(1 - \frac{\xi^2}{\xi_f^2}\right) d\xi = 1. \tag{2.22}$$

We conclude that $\xi_f = \sqrt[3]{9}$, and the final form of the solution appears to be very simple:

$$H = \begin{cases} \dfrac{\sqrt[3]{3}}{4} \left(\dfrac{I^2}{\kappa t}\right)^{1/3} \left(1 - \dfrac{x^2}{(9I\kappa t)^{2/3}}\right), & 0 \le |x| \le x_f, \\[2mm] 0, & |x| \ge x_f, \end{cases} \tag{2.23}$$

where the coordinate of the extending water front is given by

$$x_f = (9I\kappa t)^{1/3}. \tag{2.24}$$

It is of importance that in contrast with the linear equation of heat conduction one has here a finite speed of propagation: the perturbation zone $-x_f(t) \le x \le x_f(t)$ is bounded for any finite time.[2] The solution (2.23) belongs to a more general class of solutions obtained by Zeldovich and Kompaneets (1950) and Barenblatt (1952). Later the solutions of this class were obtained also by Pattle (1959).

---

[2] We consider here for comparison the solution, of concentrated-instantaneous-source type, for the *linear* equation

$$\partial_t H = \kappa \partial_{xx}^2 H \tag{I}$$

satisfying the same initial condition (2.7) and $H_i = 0$, $\ell = 0$. The solution depends on the parameters t, the integral head I determined by (2.6), x and κ. Their dimensions in this case are as follows:

$$[H] = \frac{M}{LT^2}, \qquad [t] = T, \qquad [I] = \frac{M}{T^2}, \qquad [x] = L, \qquad \kappa = \frac{L^2}{T}. \tag{II}$$

Dimensional analysis gives in the same way as previously

$$H = \frac{I}{\sqrt{\kappa t}} \Phi(\xi), \qquad \xi = \frac{x}{\sqrt{\kappa t}}. \tag{III}$$

Substituting (*III*) into equation (*I*) we obtain for the function Φ an ordinary differential equation,

$$\frac{d^2 \Phi}{d\xi^2} + \frac{\xi}{2} \frac{d\Phi}{d\xi} + \frac{\Phi}{2} = 0. \tag{IV}$$

Integration of this equation along the same lines as (2.14) gives

$$\Phi = \frac{1}{2\sqrt{\pi}} e^{-\xi^2/4}, \tag{V}$$

so that the well-known solution

$$H = \frac{I}{2\sqrt{\pi \kappa t}} e^{-x^2/(4\kappa t)} \tag{VI}$$

is obtained. Thus, we see that in the linear case the perturbation is different from zero at arbitrarily large x for arbitrary small times – an infinite speed of perturbation propagation.

In fact, the solution (2.23) is not, as we have already mentioned, a classical solution to the differential equation (2.3). Indeed, equation (2.3) contains a space derivative of the second order; meanwhile even the first space derivative of solution (2.23) is discontinuous. Therefore an important mathematical question appeared when solutions of the type (2.23) were first obtained: in what sense is (2.23) a solution of the partial differential equation (2.3) and is it unique? These questions were answered in the paper by Oleynik, Kalashnikov and Chzhou Yui-lin (1958). They introduced the natural class of so-called weak (generalized) solutions of the equations of the type (2.3) and proved the existence and uniqueness of such solutions. A very special property of solution (2.23) is, as mentioned above, the finite speed of propagation. This property was rigorously proved in the above-mentioned paper (note also an earlier paper, Barenblatt and Vishik 1956). Later, investigation of the solutions to equations of the type (2.3), now known in mathematical literature as the *porous-medium equations*, became the subject of research for many mathematicians. The fundamental reviews by Kalashnikov (1987) and Aronson (1986) are highly recommended in this respect.

The solution (2.23) is a self-similar one: there exist time-dependent scales of the head and length,

$$H_0(t) = H(0, t) = \frac{\sqrt[3]{3}}{4} \left(\frac{I^2}{\kappa t} \right)^{1/3}, \qquad x_f(t) = (9I\kappa t)^{1/3},$$

such that the head distribution at various instants can be represented in the form

$$H = H_0(t) f \left(\frac{x}{x_f(t)} \right). \tag{2.25}$$

Hence it follows that if we describe this distribution in reduced, self-similar, coordinates $H/H_0(t)$, $x/x_f(t)$ then the head distributions for any value of time are represented by a single curve.

The same situation happens for G.I. Taylor's self-similar solution to the problem of a very intense explosion, considered in the Introduction: the distributions of every property u can be represented in the self-similar form

$$u = u_0(t) f \left(\frac{r}{r_f(t)}, \gamma \right) \tag{2.26}$$

where γ is a constant adiabatic index, $r_f(t) = C(\gamma)(Et^2/\rho_0)^{1/5}$, $u_0(t) = \rho_0$ for the density, $u_0(t) = \rho_0^{3/5} E^{2/5} t^{-6/5}$ for the pressure and $u_0(t) = (Et^{-3}/\rho_0)^{1/5}$ for the velocity.

Self-similar solutions are encountered in many branches of mathematical physics. Obtaining a self-similar solution has always been regarded as a success by researchers. The basic point is that in many cases self-similarity allows one to reduce a problem involving partial differential equations (which are frequently nonlinear, so that this is especially important) to one involving ordinary differential equations. According to the hierarchy of difficulties that existed in the pre-computer era, this made certain studies easier to carry out. Moreover, self-similar solutions have been widely used as standards for evaluating all kinds of approximation method, irrespective of the immediate urgency of the problems described by the self-similar solutions themselves. The appearance of computers changed the attitude towards self-similar solutions but did not reduce the need for, and interest in, them: self-similarity continued to attract even more attention than before, but now as a deep physical property of a process which indicates that it stabilizes itself in a certain way. The statement that a phenomenon has stabilized or is entering a steady-state regime is, clearly, highly informative. The statement that a phenomenon is entering a self-similar regime is every bit as informative.

Self-similar solutions are always solutions to limiting, 'idealized', problems for which the parameters having the same dimensions as the independent variables are equal to zero or infinity. Thus in the very-intense-explosion problem the initial energy was assumed to be concentrated in a point, the explosion was assumed to be instantaneous and the initial air pressure in the ambient atmosphere to be negligibly small. Also, in the concentrated flooding problem considered above the horizontal extent of the initial dome was considered to be equal to zero and the initial water head in the stratum to be negligibly small. If this had not been the case, self-similarity would not have existed; see the relation (2.12). Therefore, for a long time self-similar solutions were treated by most researchers as though they were merely isolated 'exact special solutions' to very special problems, elegant, sometimes useful but extremely limited in significance.

It was only gradually realized that these solutions were actually of much broader significance. In fact, self-similar solutions turn out not only to describe the behaviour of physical systems under some special conditions but also to describe the 'intermediate-asymptotic' behaviour of solutions to broader classes of problems, i.e. the behaviour in the regions where these solutions have ceased to depend on the fine details of the initial conditions or boundary conditions but where the system is still far from its final equilibrium state. This situation is common, and it greatly increases the significance of self-similar solutions.

2.3 The intermediate asymptotics

We will now give an exact definition of intermediate asymptotics. As a reminder, an asymptotic representation, or an *asymptotics* for short, is an approximate representation of a function valid in a certain range of independent variables. The intermediate asymptotics for a phenomenon is determined in the following way. Assume that in the phenomenon under consideration there exist two values of an independent variable x, x_1 and x_2, having widely different magnitudes:[3]

$$x_1 \lll x_2.$$

Then the asymptotic representation of certain properties of the phenomenon in the range

$$x_1 \lll x \lll x_2$$

corresponding to values of the independent variable x that are large in comparison with the first scale x_1 but small in comparison with the second scale x_2 is called the *intermediate asymptotics*. More precisely, if in a problem we have two widely different scales x_1 and x_2 in the values an independent variable x then we call the intermediate asymptotics an asymptotic representation for $x/x_1 \to \infty$ but $x/x_2 \to 0$.

As mentioned earlier, self-similar solutions are of intrinsic interest not only, and not mainly, as exact solutions of isolated, albeit urgent, specific problems but above all as intermediate-asymptotic representations of the solutions to much wider classes of problems. We will demonstrate this point now for the problem of groundwater dome evolution after very intense concentrated flooding, presented in the previous section. In fact, the flooding is concentrated not in a single section $x = 0$ of the stratum, as was assumed in the ideal-problem formulation, but in a section of finite width 2ℓ. Also, the initial water head in the stratum H_i is not equal to zero. Therefore in addition to the dimensionless argument $\Pi_1 = \xi = x/(I\kappa t)^{1/3}$, two other arguments appeared for the function Φ in (2.12):

$$\Pi_2 = \frac{\ell}{(I\kappa t)^{1/3}}, \qquad \Pi_3 = \frac{H_i}{I^{2/3}\kappa^{-1/3}t^{-1/3}}. \qquad (2.27)$$

It is intuitively clear that the fine details of the initial dome shape and its formation have an effect only at the early stage when the dome is spread to a distance of the order of its initial width ℓ. We will abandon the consideration of such details, i.e. we shall be interested in the dome spreading only when the fluid front has travelled distances large in comparison with ℓ. This means that $x_f \gg \ell$. However, according to (2.24) x_f is of the order of $(I\kappa t)^{1/3}$; it follows that here we must have $t \gg \ell^3/(I\kappa)$. For these values of t the parameter Π_2 is much

---

[3] The symbol $a \lll b$ means that there exists an interval in the values of a variable y such that $y \gg a$ but at the same time $y \ll b$.

smaller than unity. One ordinarily assumes that if some similarity parameter has a value much smaller or much larger than unity then the dependence on that parameter, and consequently also on the corresponding dimensional parameter can be neglected. In the present case this turns out to be correct (see the next chapter), so that for $t \gg \ell^3/(I\kappa) = T_1$ the dependence of the solution on the dimensionless parameter $\Pi_2 = \ell/(I\kappa t)^{1/3}$, and therefore on the dimensional parameter ℓ, is unimportant.

Further, since the flooding is very intense, the head in the region traversed by the dome is at first very large, much greater than the initial water head in the stratum H_i. The head in the dome is of the order of $I^{2/3}\kappa^{-1/3}t^{-1/3}$, therefore at $t \ll I^2/(\kappa H_i^3) = T_2$ the parameter $\Pi_3 \ll 1$ and the initial head can be considered (the same logic!) as unimportant. Keeping in mind that for such t values $x_f \ll I/H_i$, we find that for sufficiently intense and sufficiently concentrated flooding (large I and small ℓ) the characteristic upper and lower time scales of the problem,

$$T_1 = \frac{\ell^3}{I\kappa}, \qquad T_2 = \frac{I^2}{\kappa H_i^3}, \tag{2.28}$$

and the upper and lower length scales of the problem,

$$L_1 = \ell, \qquad L_2 = \frac{I}{H_i}, \tag{2.29}$$

are strongly separated from each other, i.e. they are such that $T_1 \lll T_2$ and $L_1 \lll L_2$. The self-similar solution (2.23) describes the phenomenon of very intense and concentrated flooding at times and distances from the center of flooding large enough to make the influence of the fine details of flooding, including the initial width of the dome, disappear and at the same time small enough to make the initial groundwater head in the stratum negligible:

$$\frac{\ell^3}{I\kappa} \ll t \ll \frac{I^2}{\kappa H_i^3}$$
$$\ell \ll x \ll \frac{I}{H_i}. \tag{2.30}$$

The situation is analogous to the problem of very intense blast waves formed after an atomic explosion. In the latter case we must take into consideration that the energy release occurs not at a point but in a sphere of radius r_0 (r_0 corresponds to the time when the intense shock wave outstrips the thermal wave) and that outside this sphere the ambient gas of density ρ_0 is under a pressure that is not zero but has some finite value p_0. The solution discussed in the Introduction represents an intermediate asymptotics for

$$T_1 = \left(\frac{\rho_0 r_0^5}{E}\right)^{1/2} \ll t \ll \left(\frac{\rho_0 E^{2/3}}{p_0^{5/3}}\right)^{1/2} = T_2,$$

$$L_1 = r_0 \ll r \ll \left(\frac{E}{p_0}\right)^{1/3} = L_2,$$

(2.31)

i.e. for times, and at distances from the centre of the explosion, that are sufficiently large that the influence of the size of the initial energy discharge disappears and at the same time sufficiently small that the influence of the counter-pressure p_0 is not yet felt.[4] It is recommended that the reader should check the details of this example as an exercise: the discussion is completely analogous.

These examples clarify the basic idea of a self-similar solution as an intermediate asymptotics. The idea has been widespread that obtaining self-similar solutions is always connected with dimensional analysis, so that by applying dimensional analysis to the formulation of an idealized problem that has self-similar solutions one can always obtain the form of the solutions, i.e. a relation for the self-similar variables. Then, after obtaining an exact solution it should be easy to find a class of non-idealized problems for which the self-similar solution considered is an intermediate asymptotics. In fact, this is actually the situation for *some* self-similar solutions: we have considered examples and have indicated a general approach which is applicable in such cases.

As a rule, however, the situation is different: there exist extensive classes of problems for which a self-similar intermediate asymptotics exists but cannot be obtained from the original formulation of the problem by dimensional analysis. The form of self-similar variables, generally speaking, is obtained from the solution of certain nonlinear eigenvalue problems. It is not a question here of rare exceptions but rather of the rule: the set of self-similar solutions which cannot be obtained from dimensional considerations is considerably richer than the set of self-similar solutions whose form is completely determined by dimensional analysis.

---

[4] An historic note: Ya.B. Zeldovich once told the author that, sometime in the early 1940s, L.D. Landau showed him (long before anything about this subject was published) the asymptotic solution to gas dynamics equations similar to one presented in the Introduction. Zeldovich persuaded Landau that this solution had no terrestrial applications (perhaps only some astrophysical ones?) because a ball of the strongest explosive known at that time, tri-nitrotoluol, would be too large for there to be any niche for such a solution; using modern language, there would be no interval for intermediate asymptotics. Traces of this discussion can be found in an early book of Zeldovich (1946), where the 'Landau asymptotics' is mentioned. Very soon both Zeldovich and Landau started to work in the Soviet analog of the Manhattan project and their views definitely changed.

Subsequent examination will clarify the situation here. Modifying the flooding example considered above in such a way that at first glance all dimensional considerations used, and hence also everything deduced from them, must remain valid, we will arrive at a contradiction. Resolving the contradiction will lead us to a new class of self-similar solutions; we will call them *self-similar solutions of the second kind*.

The concept of intermediate asymptotics was formally introduced into mathematical physics by Ya.B. Zeldovich and the present author (see Barenblatt and Zeldovich 1971, 1972; Barenblatt 1959; Zeldovich and Raizer 1966, 1967) although it was implicitly used long before. In fact, this concept has important general significance. For instance, it is always used in our perception of visual art; see, for example, the painting by Salvador Dali, reproduced as the frontispiece. This painting has an instructive story. In *Scientific American*, in 1973, there appeared an article 'The recognition of faces' by Leon Harmon, a computer scientist then working at IBM. In this article several pictures were represented by sets of tiles, each one of which was a monochromatic square. The first example used 560 such squares to represent, in color, Leonardo's Mona Lisa. I used this picture in an earlier book (Barenblatt 1996) as an instructive example, illustrating the idea of intermediate asymptotics. But Harmon's article contained another example: a tiled black-and-white picture of Lincoln (see Figure 2.3) composed of 252 monochromatic squares. Its basis is widely known: the 1864 photograph used to make the $5 bill. And this same tiled picture inspired Salvador Dali to create the painting shown in the frontispiece, where he used other tiles for different images, including Harmon's original picture and paintings of his wife Gala. This painting is an especially remarkable illustration of the concept of inermediate asymptotics because it is also multiscale!

Indeed, generally we look at paintings from distances large enough not to see the brush-strokes but at the same time small enough that we can enjoy not only the painting as a whole but also its important details. Remember also *Gulliver's Travels* by Jonathan Swift. Gulliver's impressions of the fine details of the skin of a giant Brobdingnag beauty, who had the custom of putting him upon her breast, are especially instructive. It is clear from Gulliver's description that her admirers restricted themselves to an intermediate asymptotic perception. Also, the idea of intermediate asymptotics was well presented by the Russian poet Alexander Block in his poem 'The retribution':

> Rule out the accidental features
> And you will see: the world is marvellous

(this stanza was translated from the Russian by Sir James Lighthill).

It is the same thing in any scientific study. The primary thing in which the investigator is interested is the development of the phenomenon for intermediate

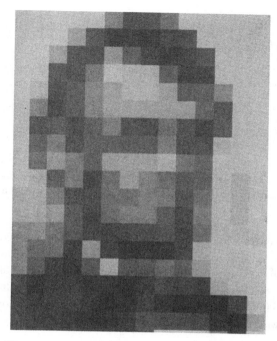

Figure 2.3. The photograph of Abraham Lincoln on a $5 bill represented (Harmon 1973) by $14 \times 18 = 252$ tiles (monochromatic squares) is an example of an intermediate asymptotics. Thus, at some intermediate distance Lincoln's portrait is easily recognizable. Close up, Lincoln's image disappears and at large distances the image becomes a blur – in effect it has disappeared again.

times and distances away from the boundaries such that the effects of accidental features or fine details in the spatial structure of the boundaries have disappeared but the system is still far from its ultimate equilibrium state. This is precisely where the underlying laws governing the phenomenon appear most clearly; therefore, a chosen intermediate asymptotics is of primary interest in every scientific study.

2.4 Problem: very intense groundwater pulse flow – the self-similar intermediate-asymptotic solution

Consider the groundwater flow at a bank of a river or channel after a short intense surge (or the penetration of a dam separating a reservoir of liquid waste). The bank is considered as a horizontal porous stratum lying on an impermeable bed, its horizontal extent being large (Figure 2.4). At the vertical boundary $x = 0$ the

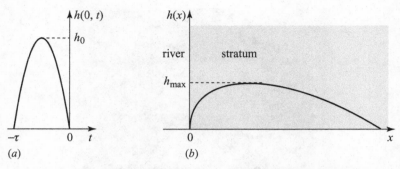

Figure 2.4. (*a*) The groundwater level at the boundary $x = 0$ as a function of time. (*b*) The schematic shape of the groundwater dome in the river bank at an arbitrary time $t > 0$.

river, channel or reservoir contacts the stratum, which is assumed to be semi-infinite, $0 \leq x < \infty$. The problem is schematically formulated as follows. At the initial moment, which we select as $t = -\tau$, the water level at the vertical boundary $x = 0$ starts to grow and quickly reaches a maximum level h_0 much larger than the initial groundwater level in the stratum. After a short time τ, i.e. at $t = 0$, the water level at the boundary $x = 0$ has returned to its initial value.

Solution. We apply to the subsequent flow equation (2.3) for the water head $H = \rho g h$. The boundary condition at the vertical boundary of the stratum $x = 0$ takes the form

$$H(0, t) = \rho g h_0 f(t/\tau) = \rho g h_0 f(\theta). \tag{2.32}$$

Here the function $f(\theta)$, $\theta = t/\tau$, is a non-dimensional function of its non-dimensional argument; it is equal to zero at $\theta = -1$, is non-negative at $-1 < \theta < 0$, reaching a maximum at a certain value of θ in this range and is equal to zero at $\theta \geq 0$.

As before we consider the initial water head in the stratum to be negligible and so assume an initial condition at $t = -\tau$ of the form

$$H(x, -\tau) \equiv 0, \qquad 0 \leq x < \infty. \tag{2.33}$$

An accurate description of the function $f(\theta)$ is in fact not needed, because we are interested in the flow at times t that are large enough that $t \gg \tau$ but not so large that the water tongue has reached the outer boundary of the stratum.

Multiplying equation (2.3) by x and integrating from $x = 0$ to $x = \infty$ (in fact, to the water front x_f), we obtain using the boundary conditions that the 'dipole moment' of the water head distribution,

$$J(t) = \int_0^{x_f(t)} x H(x, t) \, dx, \tag{2.34}$$

remains invariant for $t > 0$, so that $J(t) = J(0) = J$, where

$$J(0) = \int_0^{x_f(0)} x H(x, 0) \, dx. \tag{2.35}$$

The functions $H(x, 0)$ and $x_f(0)$ are determined by the fluid inflow into the stratum during the surge time $-\tau \leq t \leq 0$.

The intermediate asymptotics of the solution is determined as before by the the time t and the constant κ entering the basic equation (2.3). However, instead of the dome weight I, which in this case is not conserved at $t > 0$ due to the outflow of the fluid through the boundary $x = 0$, the solution is determined by the initial 'dipole moment' of the water head distribution J, which is conserved and whose dimension is obviously LMT^{-2}.

Arguments completely analogous to the derivation of the solution (2.23), (2.24) show that the self-similar intermediate-asymptotic solution takes in this case the form

$$H = \left(\frac{J}{\kappa t}\right)^{1/2} \Phi(\zeta), \qquad \zeta = \frac{x}{x_f(t)}, \tag{2.36}$$

where

$$\Phi(\zeta) = \begin{cases} \dfrac{\sqrt{5}}{3} \zeta^{1/2} (1 - \zeta^{3/2}), & 0 \leq \zeta \leq 1 \\ 0, & \zeta \geq 1. \end{cases} \tag{2.37}$$

Here $x_f(t)$ is the length of the water tongue, the horizontal extent of the groundwater dome, and is equal to

$$x_f(t) = 2(5 J \kappa t)^{1/4}. \tag{2.38}$$

The solution (2.36), (2.37) is presented in Figure 2.5 in reduced coordinates $x/x_f(t)$, $H(x, t)/H_{max}(t)$. Here $H_{max}(t) = \rho g h_{max}(t)$; $h_{max}(t)$ is the maximum height of the groundwater dome at time t. This solution was obtained by Barenblatt and Zeldovich (1957); see also Zeldovich and Raizer (1967) and Barenblatt *et al.* (1990).

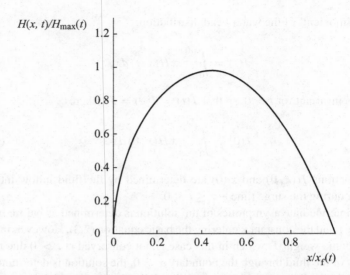

Figure 2.5. The self-similar groundwater dipole solution in reduced coordinates.

It will be useful for the reader to show that the solution (2.36) is an intermediate asymptotics of the complete solution at

$$T_1 = \tau \ll t \ll T_2 = \frac{J}{\kappa H_i^2},$$

$$L_1 = (J\kappa\tau)^{1/4} \ll x < L_2 = \left(\frac{J}{H_i}\right)^{1/2}. \tag{2.39}$$

In this intermediate time–space interval the formation details of the initial head
distribution do not play an essential role, and at the same time the maximum
value of the water head in the water dome still exceeds substantially the initial
water head in the stratum.

Chapter 3

Scaling laws and self-similar solutions that cannot be obtained by dimensional analysis

To obtain the scaling law for very intense blast wave propagation,

$$r_{\mathrm{f}} = C(\gamma)\left(\frac{Et^2}{\rho_0}\right)^{1/5},$$

G.I. Taylor, as demonstrated in the Introduction, had to replace the actual problem by an idealized one, that of a point explosion in a gas under zero pressure. In the problem of groundwater mound evolution considered in Chapter 2 we followed exactly the same method: we replaced the actual problem by an idealized one, that of infinitely concentrated instantaneous flooding with zero initial water head in the stratum. For this idealized problem we were able to demonstrate using dimensional analysis alone that the solution is self-similar and to determine the self-similar variables, i.e. the scaling laws. After obtaining the complete solution to the idealized problem we demonstrated that this self-similar solution, like the solution to the idealized problem of a very intense explosion, is an intermediate asymptotics of the solutions to a much more general class of problems.

However, the situation when everything including scaling laws can be obtained by dimensional analysis alone is in fact very rare. As a rule it turns out that, although a problem may possess a self-similar solution and scaling laws, dimensional analysis alone is insufficient to prove self-similarity starting from an idealized formulation and to find the scaling laws. We will demonstrate this more general case by modifying, seemingly slightly but in fact very essentially, the problem of groundwater dome spreading.

3.1 Formulation of the modified groundwater flow problem

Assume now that during the motion of a groundwater dome some part of the fluid is *absorbed* by the porous medium, for example by capillary imbibition.

It is easy to see that the basic balance equation (2.3) should be replaced in this case by the equation

$$m\partial_t h = \frac{k}{\mu\rho g}\, \partial_x(H\partial_x H) + Q. \tag{3.1}$$

Here Q is the specific inflow or outflow rate, i.e. the volume of fluid generated or absorbed per unit time per unit bulk volume of the medium.

Assume further that the quantity Q is proportional to the material derivative of the fluid volume between x and $x + dx$ (Figure 2.2):

$$Q = \alpha\,\frac{d(mh)}{dt}, \tag{3.2}$$

where α is a small constant. The material derivative of h is given by

$$\frac{dh}{dt} = \partial_t h + u\partial_x h,$$

where we emphasize that the fluid velocity u is different from the filtration velocity (the flux per unit of the total cross-sectional area). Our next modification of the model takes into account that the rock is *fissurized*. This means that in the rock there exists a network of fissures, or cracks, separating porous blocks. The fissures occupy a much smaller part of the total volume of the rock than the pores, so that the 'fissure porosity' m_f is much less than the porosity of the blocks: $m_f \ll m$. However, the fissures are much wider than the pores, so that the fluid is contained in the porous blocks but flows between them mainly in the fissures. Therefore the fluid velocity is determined by the relation

$$u = -\frac{1}{m_f}\frac{k}{\mu}\, \partial_x H \tag{3.3}$$

rather than by (2.2) as it was for purely porous rocks. We obtain

$$\begin{aligned}
Q &= \alpha\,\frac{d(mh)}{dt} \\
&= \alpha m\partial_t h - \frac{k}{\mu}\alpha\,\frac{m}{m_f}(\partial_x H)(\partial_x h) \\
&= \frac{1}{\rho g}\left[\alpha m\partial_t H - \frac{k}{\mu}\alpha\,\frac{m}{m_f}(\partial_x H)^2\right].
\end{aligned} \tag{3.4}$$

Substituting (3.4) into (3.1) we obtain the modified equation

$$m(1 - \alpha)\partial_t H = \frac{k}{\mu}\left[H\partial_{xx}^2 H + \left(1 - \alpha\,\frac{m}{m_f}\right)(\partial_x H)^2\right].$$

Bearing in mind that α is small we neglect it on the left-hand side. On the right-hand side α is multiplied by a large number, m/m_f, the ratio of the block porosity to the fissure porosity. Therefore the term $\alpha m / m_f = c$ on the right-hand side remains, and we come to the modified equation for the groundwater head $H(x, t)$, the *filtration–absorption equation*:

$$\partial_t H = 2\kappa [H \partial^2_{xx} H + (1 - c)(\partial_x H)^2]. \tag{3.5}$$

Here the constant κ is, as before, determined by (2.4): $\kappa = k/(2m\mu)$. When there is no absorption, $c = 0$ and we return to equation (2.3). The filtration–absorption equation (3.5) has been the subject of deep mathematical investigation (see Bertsch, Dal Passo and Ughi 1986, 1992 and the references therein).

We will try now to solve for the modified equation (3.5) the same idealized problem, of very intense concentrated flooding. In this case, to account for the modification of (2.3) the dimensionless constant c should be added to the arguments of the function f in relation (2.8); this obviously cannot change anything in an application of dimensional analysis.

3.2 Direct application of dimensional analysis to the modified problem

Repeating exactly the same arguments as in the case of no absorption, we obtain instead of (2.12) the following expression for the groundwater head:

$$H = \frac{I^{2/3}}{(\kappa t)^{1/3}} \; \Phi \left(\frac{x}{(I\kappa t)^{1/3}}, \frac{\ell}{(I\kappa t)^{1/3}}, \frac{H_i}{I^{2/3}(\kappa t)^{-1/3}}, c \right). \tag{3.6}$$

The relation (3.6) within the frames of the proposed problem formulation is indisputable; as yet we have made no assumptions. Now let us try to do exactly as we did in the previous chapter for the case of no absorption. We replace the problem by an idealized one, i.e. we assume that

1. the dome is concentrated initially at $x = 0$, i.e. $\ell = 0$, and
2. the initial groundwater head in the stratum is negligible, $H_i = 0$.

We obtain for the head $H(x, t)$ and the water-front position $x_f(t)$,

$$H = \frac{I^{2/3}}{(\kappa t)^{1/3}} \; \Phi(\xi, 0, 0, c) = \frac{I^{2/3}}{(\kappa t)^{1/3}} \; F(\xi, c),$$

$$\xi = \Pi_1 = \frac{x}{(I\kappa t)^{1/3}}, \qquad x_f = \xi_f(c)(I\kappa t)^{1/3}. \tag{3.7}$$

Relations (3.7) show readily that for $c > 0$ something is incorrect in our arguments. Indeed, integrating relation (3.5) with respect to x from $-x_f$ to x_f and using the condition of no flux, $H \partial_x H = 0$, at the water fronts $x = -x_f$ and

$x = x_f$ (the motion is symmetric), we obtain

$$\frac{d}{dt} \int_{-x_f}^{x_f} H(x,t)\,dx = -2\kappa c \int_{-x_f}^{x_f} (\partial_x H)^2 dx < 0. \tag{3.8}$$

Meanwhile according to (3.7) the dome weight $I(t)$ should be constant:

$$I(t) = \int_{-x_f}^{x_f} H(x,t)\,dx = \frac{I^{2/3}}{(\kappa t)^{1/3}}(I\kappa t)^{1/3} \int_{-\xi_f}^{\xi_f} F(\xi,c)\,d\xi = \text{const } I, \tag{3.9}$$

so that $dI(t)/dt$ should be equal to zero.

This contradiction shows that a self-similar solution to the equation (3.5) in the form (3.7) at $c > 0$ does not exist.

3.3 Numerical experiment. Self-similar intermediate asymptotics

Thus, seemingly there appears a paradox. Using exactly the same arguments which led us to a simple solution (2.23) for the case of no absorption, $c = 0$, we came to the conclusion that for $c > 0$ the solution predicted by dimensional analysis does not exist.

To understand what has happened let us turn to the results of numerical experiment. The initial condition

$$H(x,0) = \frac{I}{\ell} f_0\left(\frac{x}{\ell}\right), \qquad I = \int_{-\ell}^{\ell} H(x,0)\,dx \tag{3.10}$$

for $\ell = 0$ is represented by a generalized function, a delta function, and this cannot be introduced into the computer code. Therefore the initial condition (3.10) is used, assuming ℓ to be finite. The space coordinate x can be renormalized as well as the time t; therefore we can take $\ell = 1$ in the numerical experiments, and thus the function $f_0(\zeta)$ is assumed to be represented as a rectangle slightly 'smoothed' at the edges:

$$f_0(\zeta) = \begin{cases} 1, & |\zeta| \le 1, \\ 0, & |\zeta| > 1. \end{cases} \tag{3.11}$$

This means that $I = 2$, and the value $1/2$ was taken for κ. The numerical experiments were performed by Dr A.E. Chertock for the values $c = 0$, $c = 1/2$, $c = 2/3$, and $c = 3/4$.

The results of the experiments (see Figures 3.1–3.3) are instructive. For all values of c it was found that $H(0,t)$ and $x_f(t)$ tend at large times to scaling laws:

$$H(0,t) = A(\kappa t)^{-\lambda}, \qquad x_f(t) = B(\kappa t)^{\mu}. \tag{3.12}$$

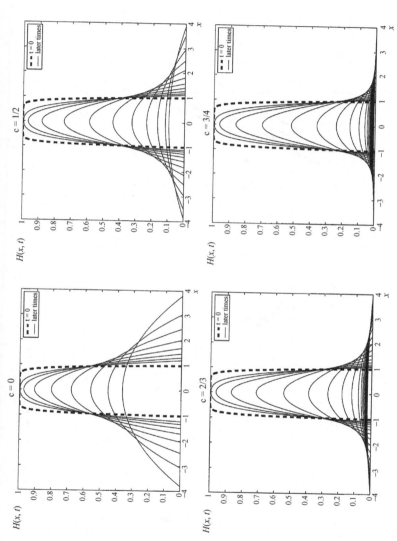

Figure 3.1. Time sequences for water head $H(x, t)$ in the natural coordinate x, for various values of c: (a) $c = 0$, (b) $c = 1/2$, (c) $c = 2/3$, (d) $c = 3/4$; $\blacksquare\blacksquare$, $t = 0$; ——, later times.

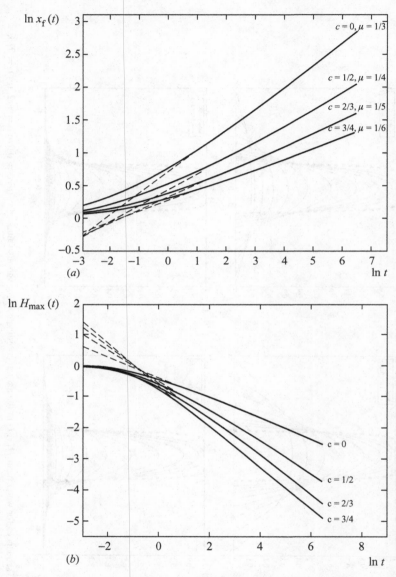

Figure 3.2. (a) $\ln x_{\mathrm{f}}(t)$ and (b) $\ln H_{\max}(t)$ plotted against $\ln t$ for various values of c. In the case $c = 0$ the asymptotic straight lines give the predicted values $\lambda = -1/3$, $\mu = 1/3$. In all other cases the values λ, μ are different from the values predicted by dimensional analysis.

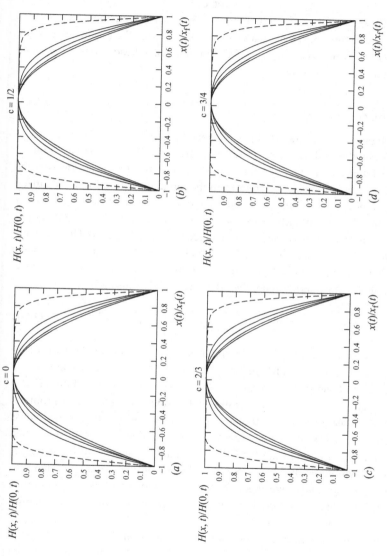

Figure 3.3. Time sequences for the water head $H(x, t)$ plotted in reduced coordinates $H(x, t)/H(0, t)$ against $x/x_f(t)$ for various values of c: (a) $c = 0$, (b) $c = 1/2$, (c) $c = 2/3$, (d) $c = 3/4$; - - -, $t = 0$, ——, later times. It is seen that with increasing t the distributions in the reduced coordinates collapse for all values of c to the same parabola, $H(x, t)/H(0, t) = 1 - x^2/x_f^2$.

However (see Figures 3.2), only for the case of no absorption, $c = 0$, do the calculated values λ and μ correspond to the predictions of dimensional analysis, $\lambda = 1/3$, $\mu = 1/3$. In other cases the following values were obtained with high accuracy: $\lambda = 1/2$, $\mu = 1/4$ for $c = 1/2$; $\lambda = 3/5$, $\mu = 1/5$ for $c = 2/3$; $\lambda = 2/3$, $\mu = 1/6$ for $c = 3/4$. Moreover, the computations showed (see Figures 3.3) that, in all cases, in the 'reduced' coordinates $x/x_f(t)$, $H(x, t)/H(0, t)$ the distributions of the groundwater head tend to the same parabola as in the case $c = 0$:

$$\frac{H(x, t)}{H(0, t)} = 1 - \frac{x^2}{x_f^2}, \quad |x| < x_f; \qquad H(x, t) \equiv 0, \quad |x| > x_f. \tag{3.13}$$

We have to explain now what has happened and especially why the asymptotics (3.12), the scaling laws, are different from the predictions based on dimensional analysis. In fact, the initial condition (3.10), (3.11) assumed in the numerical computations is different from the initial condition for the idealized problem: a new dimensional parameter ℓ, the initial groundwater dome extension, is introduced explicitly. This parameter was absent in the formulation of the idealized problem because it was assumed to be equal to zero. However, the condition (3.10) with $\ell = 0$ and $I > 0$ cannot be introduced into the computer code. Therefore the standard procedure of dimensional analysis gives for the problem under consideration not the relations (3.7) but instead:

$$H = \frac{I^{2/3}}{(\kappa t)^{1/3}} \Phi(\xi, \eta, c), \qquad \xi = \frac{x}{(I\kappa t)^{1/3}}, \qquad \eta = \frac{\ell}{(I\kappa t)^{1/3}}; \tag{3.14}$$

as before we assume $H_i = 0$. This relation is also valid for the case of no absorption, $c = 0$: the self-similar solution (2.23) then corresponds to the singular initial condition (3.10) at $I > 0$, $\ell = 0$. However, the solution (2.23) is not merely an exact special solution to this singular initial-value problem. Indeed, let us assume for $c = 0$ that ℓ is finite. At $t \to \infty$ the dimensionless parameter $\eta = \ell/(I\kappa t)^{1/3}$ tends to zero. At the same time we can always adjust x in such a way that $\xi = x/(I\kappa t)^{1/3}$ will remain constant with increasing t. Therefore asymptotically, at $t \to \infty$, the solution for $c = 0$ and $\ell > 0$ tends to the solution (2.23), and so we conclude that the solution (2.23) is an asymptotics for a class of initial-value problems, not just a special solution corresponding to specific singular initial data.

This is so because at $c = 0$ the limit of the function $\Phi(\xi, \eta, 0)$ as $\eta \to 0$ exists and is different from zero. The non-existence of a solution to the problem with singular initial data for finite absorption, i.e. $c > 0$, means that, contrary to the case $c = 0$, for $c > 0$ the function $\Phi(\xi, \eta, c)$ entering relation (3.14) does not have a finite non-zero limit as $\eta = \ell/(I\kappa t)^{1/3} \to 0$. However, the numerical experiment shows that there is something special for $c > 0$: the self-similar asymptotics of the solution exists, although it cannot be represented in the simple form (2.13). In fact, the numerical experiment suggests that the function $\Phi(\xi, \eta, c)$ has a 'scaling' asymptotics at $\eta \to 0$ and that the scaling law exists also for the water front:

$$\Phi(\xi, \eta, c) = \text{const}_1 \ \eta^p f(\zeta, c), \qquad \zeta = \frac{x}{x_f}, \quad x_f = \text{const}_2 \ (I\kappa t)^{1/3} \eta^q. \quad (3.15)$$

Here const_1, const_2, p and q are constants; p and q are non-zero at $c > 0$. Substituting (3.15) into the solution (3.14), we obtain that at small η, i.e. at $t \gg \ell^3/(I\kappa)$, (3.14) has a self-similar asymptotics

$$H = A(\kappa t)^{-\lambda} f(\zeta, c), \qquad \zeta = \frac{x}{x_f(t)} = \frac{x}{B(\kappa t)^\mu}. \quad (3.16)$$

In (3.16) we have used the notation

$$\lambda = \frac{1}{3} + \frac{p}{3}, \qquad \mu = \frac{1}{3} - \frac{q}{3},$$
$$A = \text{const}_1 \ I^{2/3 - p/3} \ell^p, \qquad B = \text{const}_2 \ I^{1/3 - q/3} \ell^q. \quad (3.17)$$

Now substitute (3.16) into the basic equation for the groundwater head (3.5):

$$2\frac{A}{B^2(\kappa t)^{\lambda + 2\mu - 1}} \left[f\frac{d^2 f}{d\zeta^2} + (1 - c)\left(\frac{df}{d\zeta}\right)^2 \right] + \mu\zeta\frac{df}{d\zeta} + \lambda f = 0,$$
$$\zeta = \frac{x}{B(\kappa t)^\mu}. \quad (3.18)$$

The function $f(\zeta, c)$ does not depend on time explicitly, only via the variable ζ. Therefore the time exponent $\lambda + 2\mu - 1$ in (3.18) must vanish and so we obtain $\lambda = 1 - 2\mu$.

Also, we have the freedom to normalize arbitrarily the function f. Therefore the factor const_1 can be selected arbitrarily. We take $A = B^2\mu$ and come to an ordinary differential equation for $f(\zeta, c)$:

$$2\left[f\frac{d^2 f}{d\zeta^2} + (1 - c)\left(\frac{df}{d\zeta}\right)^2 \right] + \zeta\frac{df}{d\zeta} + \frac{1 - 2\mu}{\mu} f = 0. \quad (3.19)$$

Writing ξ_f for const_2 and using (3.16) and (3.17) we obtain

$$x_f(t) = \xi_f I^\mu \ell^{1 - 3\mu}(\kappa t)^\mu. \quad (3.20)$$

3.4 Self-similar limiting solution. The nonlinear eigenvalue problem

The scaling laws (3.15) observed in the numerical experiment suggest that at large times the asymptotic representation of the solution to the actual problem of concentrated but not infinitely concentrated flooding can be represented in the form

$$H = \frac{B^2\mu}{[\kappa t]^{1-2\mu}} f\left(\frac{x}{B[\kappa t]^\mu}, c\right), \qquad B = \xi_f\left(I\ell^{(1-3\mu)/\mu}\right)^\mu \qquad (3.21)$$

where the function $f(\zeta, c)$ satisfies the ordinary differential equation (3.19). The exponent μ is a priori unknown and is to be determined in the course of solution.

Owing to the natural assumption of symmetry of the asymptotic solution we can construct the solution of equation (3.19) at $\zeta \geq 0$ and assume a first boundary condition,

$$\frac{df}{d\zeta} = 0 \qquad \text{at} \quad \zeta = 0. \qquad (3.22)$$

The second obvious boundary condition comes from the continuity of the water head at the water front $x = x_f(t)$: $H(x_f(t), t) = 0$. This gives

$$f(1, c) = 0 \qquad \text{at} \quad \zeta = 1. \qquad (3.23)$$

The function $f(\zeta, c)$ must also satisfy another condition, which we will now derive. According to the Darcy law, the finiteness of the filtration velocity at the water front requires finiteness of the head gradient $\partial_x H$ at the front. Clearly this condition provides continuity of the water flux at the front, which is proportional to $H\partial_x H$. (At $x \geq x_f$ the head and flux are identically zero, as discussed earlier.) However, contrary to the case $c = 0$, in the case $c > 0$ the condition of the fluid flux continuity at the front is not sufficient for the formulation of a well-posed problem, and a stronger condition, that of the finiteness of the filtration velocity at the water front, should be imposed. To show this, substitute into equation (3.19) $f = \zeta^2\phi(\eta, c)$, where $\eta = \ln \zeta$, so that $d\eta/d\zeta = 1/\zeta$. We obtain the following equation, which does not contain an independent variable:

$$4\phi^2 + 6\phi\frac{d\phi}{d\eta} + 2\phi\frac{d^2\phi}{d\eta^2} + 8(1-c)\phi^2 + 8(1-c)\phi\frac{d\phi}{d\eta}$$

$$+ 2(1-c)\left(\frac{d\phi}{d\eta}\right)^2 + 2\phi + \frac{d\phi}{d\eta} + \frac{1-2\mu}{\mu}\phi = 0.$$

According to a common rule we take $\psi = d\phi/d\eta$ as the new unknown, and ϕ as the independent variable, so that $d^2\phi/d\eta^2 = \psi(d\psi/d\phi)$ and the previous equation is reduced to an equation of the first order,

$$\frac{d\psi}{d\phi} = -\frac{A}{2\phi\psi},$$

where

$$A = [4 + 8(1 - c)]\phi^2 + [6 + 8(1 - c)]\phi\psi + 2(1 - c)\psi^2 + \frac{1}{\mu}\phi + \psi.$$

We are interested in the behaviour of the integral curves of (3.19) at the point $\zeta = 1$, $f = 0$. Therefore the behaviour of the integral curves of the equation of the first order should be investigated at $\psi \gg \phi$, when this equation can be represented approximately in the form

$$\frac{d\psi}{d\phi} = -\frac{2(1 - c)\psi + 1}{2\phi}.$$

It is easy to show by integration of the last equation that all the integral curves of (3.19) at the point $\zeta = 1$, $f = 0$, except for one, behave like $f = C(1 - \zeta)^{1/(2-c)} +$ small quantities, where $C > 0$ is a constant parameter of the family. For all these curves the velocity of filtration at the front, proportional to $df/d\zeta$ at $\zeta = 1$, is infinite. However, the flux, proportional to $df^2/d\zeta$ at $\zeta = 1$, is equal to zero, so that at $c > 0$ there is no discontinuity of flux when matching one of these curves with $f \equiv 0$. Contrary to this, the exceptional curve, a separatrix, corresponds to the condition $2(1 - c)\psi + 1 = 0$, so that close to $\phi = 0$, $\psi = -1/2(1 - c)$, and returning to the function $f(\zeta, c)$ we obtain that the velocity of filtration at the front, proportional to $df/d\zeta$ at $\zeta = 1$, is finite and the condition holds:

$$\frac{df}{d\zeta}(1, c) = -\frac{1}{2(1 - c)}. \tag{3.24}$$

Therefore the solution corresponding to the exceptional curve should be selected.

We emphasize that the condition (3.24) does not follow automatically from equation (3.19). It required an additional physical assumption, of the finiteness of the fluid velocity at the water front. We repeat that for the classic Boussinesq equation ($c = 0$) the condition of continuity of flux at the groundwater front was enough to come to a unique solution. For $c > 0$ this is not the case, and a more subtle condition, the finiteness of the filtration velocity, is needed. This important argument is due to J.L. Vázquez.

Thus we have obtained for the *second-order* equation (3.19) *three* boundary conditions, (3.22), (3.23) and (3.24). For arbitrary values of the parameter μ a solution to the equation (3.19) satisfying these three boundary conditions does not exist. However, the parameter μ is a priori unknown and must be determined. Therefore we have to find not only the solution $f(\zeta, c)$ but also the value of the parameter μ – the *eigenvalue* – for which the solution does exist.

Thus our problem of finding the intermediate asymptotics of the water head after a concentrated flooding is reduced to a *nonlinear eigenvalue problem*: that of solving the second-order equation (3.19) with three boundary conditions (3.22), (3.23), and (3.24) and determining the eigenvalue μ – the value of the parameter entering the equation for which such a solution exists. In fact the solution to this problem is very simple. As is easy to check,

$$f(\zeta, c) = C(1 - \zeta^2), \qquad \mu = \frac{1-c}{3-2c}, \qquad (3.25)$$

where

$$C = \frac{1}{4(1-c)}. \qquad (3.26)$$

So, the intermediate-asymptotic solution for which we were searching takes the following form:

$$H = \frac{B^2}{4(3 - 2c)[\kappa t]^{1/(3-2c)}}\left[1 - \frac{x^2}{x_f^2}\right] \qquad \text{at } |x| \leq x_f,$$

$$H \equiv 0 \qquad \text{at } |x| \geq x_f, \qquad x_f = \xi_f\left[I\ell^{c/(1-c)}\kappa t\right]^{(1-c)/(3-2c)},$$

$$B = \xi_f\big(I\ell^{c/(1-c)}\big)^{(1-c)/(3-2c)}. \qquad (3.27)$$

Here we use the dimensionless constant ξ_f to replace the dimensional constant B.

We make several comments. At $c = 0$ the intermediate-asymptotic solution (3.27) gives us the solution (2.23), (2.24) to the idealized problem of infinitely concentrated flooding considered in Chapter 2. To simplify comparison we note that the function $F(\xi)$ entering equation (2.13) is related to the function $f(\zeta, 0)$, (3.25), in the following way:

$$F(\xi) = \xi_f^2\mu f\left(\frac{\xi}{\xi_f}, 0\right), \qquad \mu = \frac{1}{3}.$$

Therefore, according to (2.16),

$$F(\xi) = \frac{\xi_f^2}{3} f\left(\frac{\xi}{\xi_f}, 0\right), \qquad 0 \leq \xi \leq \xi_f,$$

$$f(\zeta) = \frac{1}{4}(1 - \zeta^2), \qquad \zeta = \frac{\xi}{\xi_f}.$$

Obviously this function satisfies the condition $f'(1) = -1/2$, obtained from (3.24) at $c = 0$. Furthermore, for $c > 0$ the solution (3.27) is self-similar; however, this solution does not correspond to an infinitely concentrated source and cannot be obtained from dimensional considerations – we had to solve a nonlinear eigenvalue problem to obtain the exponent μ entering the self-similar variables. Again, as in the classical case $c = 0$, for $c > 0$ in (3.27) we can pass to the limit $\ell \to 0$, for fixed values x and t. If in this passage to the limit the initial integral head I remains fixed then the solution (3.27) tends to zero. Equation (3.27) also shows that, for $c > 0$, to obtain the same limiting solution as was obtained for finite ℓ and $t \to \infty$ it is necessary to proceed to the limit $\ell \to 0$ with the initial head I simultaneously growing in such a way that the product $I\ell^{c/(1-c)}$ remains constant. Physically, it is transparent: when reducing ℓ we need to compensate properly for the reduction in fluid volume which takes place in reaching a larger value of x_f, equal to the previous ℓ. It is instructive to see that the way in which to compensate for this reduction cannot be obtained from dimensional analysis.

Furthermore, for the case of no absorption, $c = 0$, there exists an integral conservation law for the current dome weight $I(t)$,

$$I(t) = \int_{-x_f}^{x_f} H(x, t)\, dx = I(0) = I. \tag{3.28}$$

As we have seen, such an integral conservation law does not exist in the case of non-zero absorption, $c > 0$. However, this law is replaced by a certain asymptotic non-integral conservation law. Indeed, as already discussed, at the stage $t \gg \ell^3/(I\kappa)$ the solution is represented by the self-similar asymptotics (3.27). Therefore at this stage

$$\begin{aligned}
I(t) &= \int_{-x_f}^{x_f} H(x, t)\, dx = \frac{B^2}{4(3 - 2c)[\kappa t]^{1/(3-2c)}} \int_{-x_f}^{x_f} \left[1 - \frac{x^2}{x_f^2}\right] dx \\
&= \frac{\xi_f^3 \left(I\ell^{c/(1-c)}\right)^{3(1-c)/(3-2c)}}{3(3 - 2c)[\kappa t]^{c/(3-2c)}}.
\end{aligned}$$

Multiplying $I(t)$ by $x_f^{c/(1-c)}$ we obtain the conservation law

$$I(t) x_f^{c/(1-c)} = \frac{1}{3(3 - 2c)} \xi_f^{(3-2c)/(1-c)} I\ell^{c/(1-c)} = \text{const}, \tag{3.29}$$

valid at large times $t \gg \ell^3/(I\kappa)$. For $c > 0$ the quantity ξ_f cannot be obtained from the integral conservation law as was done for the case of no absorption, $c = 0$. It can be obtained, however, by matching (for example numerically) the pre-self-similar solution to the asymptotics (3.27). The quantity ξ_f which enters the asymptotics (3.27) and the conservation law (3.29) is also an invariant of the entire pre-self-similar solution. This means that if we take as an initial condition the head distribution $H(x, t_*)$ at an arbitrary time $t = t_*$ then ξ_f remains the same. Following Lax (1968) we call such invariant quantities *integrals*. However, in contrast to the case of no absorption, $c = 0$, the value of these integrals cannot be obtained from integral conservation laws valid during the entire process. We will call them therefore *implicit integrals*. The general idea and similar conservation laws for some other problems were suggested by G.K. Batchelor and P.F. Linden at the Fluid Mechanics Seminar at Cambridge.

Chapter 4
Complete and incomplete similarity. Self-similar solutions of the first and second kind

4.1 Complete and incomplete similarity

In Chapters 2 and 3 we considered two instructive and fundamentally different, albeit seemingly analogous, problems. In the problem of very intense, instantaneous and infinitely concentrated flooding considered in Chapter 2, following exactly the basic idea demonstrated in the Introduction for a very intense explosion, we arrived at an idealized statement of infinitely concentrated flooding. Applying to this idealized problem the standard procedure of dimensional analysis presented in Chapter 1 we were able to reveal the self-similarity of the solution, to find the self-similar variables and to obtain the solution in a simple closed form.

Deeper consideration showed, however, that this simplicity is illusory and that in making the assumption of an infinitely concentrated flooding we went, we might say, to the brink of an abyss. We demonstrated this when in Chapter 3 we modified the formulation of the problem, seemingly only slightly, by introducing fluid absorption. It would seem that in the modified formulation the same ideal problem statement would be possible and that all our dimensional reasoning would preserve its validity. However, in proceeding with the modified formulation we arrived at a contradiction. It turned out that in the modified formulation the solution to the ideal problem of very intense, instantaneous and infinitely concentrated flooding does not exist. More detailed analysis demonstrated that in trying to find a solution to the modified problem and blindly following G.I. Taylor's path in the formulation of an idealized problem we came to a dead end, because the very statement of the problem was improper. What was actually needed was not an exact solution of the simply formulated idealized problem of instantaneous infinitely concentrated flooding but the asymptotics of the solution to the non-idealized problem where the groundwater dome is initially concentrated in a section of finite extent. Naturally, the solution to the latter problem turned out to be non-self-similar. The passage to the limit as the width of the initially flooded section tends to zero led to an empty result – the solution tends to zero. Then we addressed numerical analysis. Its results demonstrated that a meaningful intermediate asymptotics exists and that it is self-similar although different from what was expected. It was revealed that this intermediate asymptotics, and not the solution to a limit problem, was precisely what we actually need. It turns out that we *cannot* consider the flooding as concentrated at a single plane section. On the contrary, when reducing the width of the section of stratum in which the flooding initially takes place, in order to obtain a proper asymptotics to the solution of the original non-self-similar problem, we must increase the amount of fluid released initially, and in such a way that a certain 'moment' of the initial distribution of the water head remains constant. It is important that the power to which the size of the flooding section appears in the expression of this 'moment' cannot be found in advance and that in principle it is impossible to determine this power from dimensional analysis. The power appears as an 'eigenvalue', which must be found in the course of construction of the self-similar asymptotics. Thus we came to the conclusion that the self-similar solutions are divided into two principally distinct types.

We will now give a formal classification of similarity rules, scaling laws and, in particular, self-similar solutions.

Recall from Chapter 1 that any physically significant relation[1] among dimensional (generally speaking) parameters,

$$a = f(a_1, \ldots, a_k, b_1, \ldots, b_m), \tag{4.1}$$

can be represented in the form

$$\Pi = \Phi(\Pi_1, \ldots, \Pi_i, \ldots, \Pi_m), \tag{4.2}$$

where the dimensionless parameters in (4.2) are defined using (see (1.20)) the expressions for the dimensions of $a, b_1, \ldots, b_m$ via the powers of the dimensions of $a_1, \ldots, a_k$:

$$\Pi = \frac{a}{a_1^p \cdots a_k^r}, \quad \cdots \quad \Pi_i = \frac{b_i}{a_1^{p_i} \cdots a_k^{r_i}}, \quad \cdots \quad i = 1, \ldots, m. \tag{4.3}$$

From relations (4.2) and (4.3) it follows that every function f which enters a physically significant relationship of the general type (4.1) has the property of

---

[1] We remind the reader that a physically significant relationship expresses a law valid for various observers, in particular observers whose units of measurement are of different magnitudes.

generalized homogeneity:

$$f(a_1, \ldots, a_k, b_1, \ldots, b_m) = a_1^p \cdots a_k^r \, \Phi \left(\frac{b_1}{a_1^{p_1} \cdots a_k^{r_1}}, \ldots, \frac{b_m}{a_1^{p_m} \cdots a_k^{r_m}} \right).$$
(4.4)

Usually, the parameters $a_1, \ldots, a_k$ designated as having independent dimensions are chosen to be those governing parameters which are definitely significant for the phenomenon under consideration.

Now consider the remaining governing parameters, $b_1, \ldots, b_m$. In a traditional argument 'on a physical level' a parameter b_i is considered as essential, i.e. actually governing the phenomenon, if the value of the corresponding dimensionless parameter Π_i is not too large and also not too small, to be specific, between about $1/10$ and 10.

Thus, if some of the dimensionless governing parameters $\Pi_{\ell+1}, \ldots, \Pi_m$ corresponding to the dimensional parameters $b_{\ell+1}, \ldots, b_m$ are small or large, it is assumed by a tacit convention that the influence of these dimensionless parameters, and consequently of the corresponding dimensional parameters, can be neglected.

Actually, this argument is valid sometimes but not always. It is valid if there exists a finite non-zero limit of the function Φ in (4.2) when the parameters $\Pi_{\ell+1}, \ldots, \Pi_m$ all go to zero or infinity while the other similarity parameters $\Pi_1, \ldots, \Pi_\ell$ remain constant. In fact even more is required: the function Φ must converge sufficiently fast to a finite non-zero limit as $\Pi_{\ell+1}, \ldots, \Pi_m$ go to zero or infinity. If these conditions are actually satisfied then, for sufficiently small or sufficiently large $\Pi_{\ell+1}, \ldots, \Pi_m$, the function Φ in (4.2) can be replaced by a function of a smaller number of arguments:

$$\Pi = \Phi_1(\Pi_1, \ldots, \Pi_\ell).$$
(4.5)

We met such a situation in the problem of very intense concentrated flooding with no absorption, considered in Chapter 2, where the finite non-zero limit of the function $\Phi(\Pi_1, \Pi_2, \Pi_3)$ at vanishing $\Pi_2 = \ell/(I\kappa t)^{1/3}$, $\Pi_3 = H_i/(I^{2/3}\kappa^{-1/3}t^{-1/3})$ did exist, and the limit happened to be a function of a single parameter $\Pi_1 = \xi = x/(I\kappa t)^{1/3}$.

In such cases we speak of the *complete similarity*, or *similarity of the first kind*, of a phenomenon in the parameters $\Pi_{\ell+1}, \ldots, \Pi_m$.

It is quite obvious that such a situation is far from being the general case. Usually, when the dimensionless governing parameters $\Pi_{\ell+1}, \ldots, \Pi_m$ go to zero or infinity the function $\Phi(\Pi_1, \ldots, \Pi_\ell, \Pi_{\ell+1}, \ldots, \Pi_m)$ does not necessarily tend to a limit, let alone a finite and non-zero one. Therefore, in general the parameters $b_{\ell+1}, \ldots b_m$ remain essential no matter how small or large the values of the corresponding dimensionless parameters $\Pi_{\ell+1}, \ldots, \Pi_m$ are. This statement is correct but trivial and non-constructive.

The fact of crucial importance is that there exists another class of phenomena, which also comprises exceptions. This class is, however, much wider than the class of complete-similarity phenomena; for phenomena in this class the function Φ entering (4.2) possesses at large or small values of $\Pi_{\ell+1}, \ldots, \Pi_m$ the property of generalized homogeneity in its own dimensionless arguments:

$$\Phi = \Pi_{\ell+1}^{\alpha_{\ell+1}} \cdots \Pi_m^{\alpha_m} \, \Phi_1 \left(\frac{\Pi_1}{\Pi_{\ell+1}^{\beta_1} \cdots \Pi_m^{\delta_1}}, \ldots, \frac{\Pi_\ell}{\Pi_{\ell+1}^{\beta_\ell} \cdots \Pi_m^{\delta_\ell}} \right) \tag{4.6}$$

where $\alpha_{\ell+1}, \ldots, \delta_\ell$ are constants. This is exactly the same form of generalized homogeneity as for the basic function f in relations (4.1) and (4.4). However, there is a fundamental difference. *The generalized homogeneity of the function f in (4.1), (4.4) follows from the general physical covariance principle, and the constants $p, \ldots, r_m$, in (4.4) are obtained by simple rules of dimensional analysis. In contrast, the generalized homogeneity of the function Φ in (4.6) is a special property of the problem under consideration.* Therefore the constants $\alpha_{\ell+1}, \ldots, \delta_\ell$ in principle cannot be obtained by using dimensional analysis: relation (4.2) is the most that dimensional analysis can give.

In such exceptional cases, relation (4.1) can be represented in the form, comparable with (4.5),

$$\Pi^* = \Phi_1(\Pi_1^*, \ldots, \Pi_\ell^*), \tag{4.7}$$

where

$$\Pi^* = \frac{\Pi}{\Pi_{\ell+1}^{\alpha_{\ell+1}} \cdots \Pi_m^{\alpha_m}}$$
$$= \frac{a}{a_1^{p-\alpha_{\ell+1}p_{\ell+1}-\cdots-\alpha_m p_m} \cdots a_k^{r-\alpha_{\ell+1}r_{\ell+1}-\cdots-\alpha_m r_m} b_{\ell+1}^{\alpha_{\ell+1}} \cdots b_m^{\alpha_m}} \tag{4.8}$$

and, for $i = 1, \ldots, l$,

$$\Pi_i^* = \frac{\Pi_i}{\Pi_{\ell+1}^{\beta_i} \cdots \Pi_m^{\delta_i}}$$
$$= \frac{b_i}{a_1^{p_i-\beta_i p_{\ell+1}-\cdots-\delta_i p_m} \cdots a_k^{r_i-\beta_i r_{\ell+1}-\cdots-\delta_i r_m} b_{\ell+1}^{\beta_i} \cdots b_m^{\delta_i}}. \tag{4.9}$$

Two special cases of the general relation (4.7) that have appeared frequently in recent current research are: (i) $\ell + 1 = m$, $\alpha_{\ell+1} = \alpha$, $\beta_1 = \beta$ and other powers entering Π_i equal to zero, so that

$$\Phi = \Pi_{\ell+1}^{\alpha} \Phi_1 \left(\frac{\Pi_1}{\Pi_{\ell+1}^{\beta}}, \Pi_2, \dots, \Pi_{\ell} \right); \tag{4.10}$$

and (ii) an even more special case, $\beta = 0$, for which

$$\Phi = \Pi_{\ell+1}^{\alpha} \Phi_1(\Pi_1, \Pi_2, \dots, \Pi_{\ell}). \tag{4.11}$$

We met the special case (4.10) in the problem considered in Chapter 3. In these special cases only one small or large governing parameter $\Pi_{\ell+1}$ violates complete similarity.

The function f which enters the basic relation (4.1) in these two special cases assumes the form

$$f = a_1^{p - \alpha p_m} \cdots a_k^{r - \alpha r_m} b_m^{\alpha}$$
$$\times \Phi \left(\frac{b_1}{a_1^{p_1 - \beta p_m} \cdots a_k^{r_1 - \beta r_m} b_m^{\beta}}, \frac{b_2}{a_1^{p_2} \cdots a_k^{r_2}}, \dots, \frac{b_{m-1}}{a_1^{p_{m-1}} \cdots a_k^{r_{m-1}}} \right), \tag{4.12}$$

(in the second case $\beta = 0$) which illustrates the general statement: the exponents cannot generally be found from dimensional analysis; the parameter b_m, which violated self-similarity, does not disappear although it enters in combination with parameters a and b_1 only.

Equation (4.6) shows that in these special cases there is a reduction in the number of arguments of the function Φ that defines the relationship in which we are interested, exactly as in the case of complete similarity. These arguments also are power monomials. However, the governing parameters $b_1, \dots, b_m$ do not disappear from the resulting relations; they remain essential and continue to influence the phenomenon no matter how small or large the corresponding similarity parameters are. Moreover, the powers in which these dimensional parameters enter the dimensionless parameters cannot be obtained from dimensional analysis. In such cases we speak of *incomplete similarity*, or *similarity of the second kind* of a phenomenon in the relevant parameters.

The conclusion at which we have arrived is entirely natural: if the values of certain dimensionless parameters Π_i are small or large then there are three possibilities.

1. The limits of the corresponding functions Φ as the Π_i tend to zero or infinity exist and are finite and non-zero. The corresponding governing parameters, the dimensional b_i or the dimensionless Π_i, can be excluded from consideration and the number of arguments of the functions Φ therefore decreases. All the similarity parameters can be determined by means of the regular procedures of dimensional analysis. This case

corresponds to complete similarity of the phenomenon in the similarity parameters Π_i.

2. No finite limits exist for the functions Φ as the Π_i tend to zero or infinity, but one of the special cases indicated above holds.[2] If so, the number of arguments of the functions Φ can be decreased, but not all the parameters Π, Π_i can be obtained from dimensional analysis and the governing parameters b_i remain essential no matter how small (or large) the corresponding similarity parameters. This case corresponds to incomplete similarity in the parameters Π_i.

3. No finite limits exist for the functions Φ as the $\Pi_i \to 0$ or ∞, and the indicated exceptions do not hold. This case corresponds to lack of similarity of the phenomenon in the parameters Π_i. It has already been remarked that, no matter how large (or small) the values of the parameters Π_i, in this case we cannot obtain a relation of the form (4.5) between power-type combinations of the governing and determined parameters that has a smaller number of arguments for the functions Φ.

The difficulty is that a priori, until we have obtained a non-self-similar solution of the complete non-idealized problem (in which case we do not need to use similarity methods), we do not know with which of these three cases 1–3 we are dealing, irrespective of whether we have an explicit mathematical formulation of the problem. Hence one can only recommend assuming in succession each of these possible situations for small (or large) similarity parameters – complete similarity, incomplete similarity, lack of similarity – and then comparing the relations obtained under each assumption with data from numerical calculations, experiments or the results of asymptotic analytic investigations. The term 'experimental asymptotics' proposed for such analysis by Professor Norman Zabusky seems to be very appropriate.

4.2 Self-similar solutions of the first and second kind

We now consider some problem in mathematical physics that describes certain phenomenon; let the quantity a be an unknown in this problem and let the

---

[2] The question can arise of why one regards as exceptional only asymptotic representations of the power-type forms (4.6), (4.8), (4.9), (4.10) and (4.11); is it impossible to factorize the function Φ by another function of Π_i, for example, $\log \Pi_i$? In fact, in the case where Φ is factorizable by $\log \Pi_i$ one no longer gets relations among power-type combinations of dimensional parameters, yet products of their powers give, upon multiplication, power-type combinations of the same form. As was proved in Chapter 1, dimensions are always power-type monomials. It can be obtained, by exactly the same argument, that a power-type asymptotics follows from the lack of characteristic distinguished values of the parameters $b_{\ell+1}, \ldots, b_m$.

quantities $a_1, \ldots, a_k, b_1, \ldots, b_m$ be the independent variables and parameters appearing in the equations and the boundary, initial and other conditions that determine solutions.

Self-similar solutions are always solutions of idealized (degenerate) problems obtained if certain parameters b_i and the dimensionless parameters Π_i corresponding to them assume zero or infinite values. They are simultaneously exact solutions of degenerate problems and asymptotic (generally intermediate-asymptotic) representations of the solutions of wider classes of non-idealized non-self-similar problems as the parameters b_i tend to zero or infinity.

It is clear that if an asymptotics is self-similar, and if the self-similar variables are power-law monomials, then one of the two special cases mentioned above, complete and incomplete similarity, must hold. Correspondingly, self-similar solutions are divided into solutions of the first and second kind.

Self-similar solutions of the first kind are obtained when passage to the limit, from a non-self-similar non-idealized problem to the corresponding self-similar idealized problem, gives complete similarity in the parameters that made the original problem non-idealized and its solution non-self-similar. Expressions for all the self-similar variables, independent as well as dependent, can be obtained here by applying dimensional analysis.

Self-similar solutions of the second kind are obtained in the case where the idealization of the original problem is such that there is incomplete similarity in the similarity parameters. Then expressions for the self-similar variables cannot in general be obtained from dimensional considerations. The parameters that make the problem non-idealized, and its solution non-self-similar, remain in the expressions for the self-similar variables.

In the direct construction of a self-similar solution of the second kind, determination of the exponents of the self-similar variables leads to a nonlinear eigenvalue problem. The constant multipliers appearing in the self-similar variables are left undetermined in the direct construction of self-similar solutions of the second kind. These constants can be found by following, for example by means of numerical calculations, the entire process of evolution of a solution of the non-idealized problem into a self-similar asymptotics.

If the constants can be found from integral conservation laws, this means that for an appropriate choice of governing parameters the problem can be reformulated and reduced to a problem of the first kind. For example, the classical problem of a very intense explosion is found to be represented as a self-similar solution of the second kind if one chooses the governing parameters of the non-idealized pre-self-similar problem inappropriately. The possibility of obtaining solutions to this problem as self-similar solutions of the first kind is connected with the choice, as a governing parameter, of the energy of the explosion; this, by virtue of the corresponding integral conservation law, does not vary with time.[3]

The problem of water-dome spreading in a porous stratum, considered in Chapter 2, has an intermediate asymptotics which is a self-similar solution of the first kind. In Chapter 3 a self-similar solution of the second kind appeared as an intermediate asymptotics for the solution to the problem of dome spreading when absorption is taken into account. Seemingly the formulations of both problems are very close, but the intermediate asymptotics are essentially different.

Complete similarity makes it possible to obtain meaningful scaling laws directly by dimensional analysis, without solving the whole problem. A classical example is the scaling law for very intense shock-wave propagation obtained by G.I. Taylor (see the Introduction). Another example: let us take the problem of water-dome spreading considered in Chapter 2 and try to find, without solving the whole problem, the law for decay of the maximum water head H_{max}. Clearly the maximum water head is at the center, $x = 0$. Assuming that the dome is concentrated initially at the section $x = 0$ and that the initial water head in the stratum is negligible, we find using dimensional analysis

$$H_{max} = f(I, t, \kappa), \qquad \Pi_{max} = \frac{H_{max}}{I^{2/3}(\kappa t)^{-1/3}} = \text{const},$$

whence

$$H_{max} = \text{const} \left(\frac{I^2}{\kappa t} \right)^{1/3} \qquad (4.13)$$

(Our calculation in Chapter 2 showed that const $= 3^{1/3}/4$). Such an argument is valid in this case because for the non-idealized problem

$$H_{max} = f(I, t, \kappa, \ell), \qquad \Pi_{max} = \Phi_{max}(\Pi_2), \qquad (4.14)$$

---

[3] In contrast, an integral conservation law valid also for the pre-self-similar stage is not a necessary property of the self-similar solution of the first kind. An elegant example illustrating this point was given by Entov (1994). The self-similar intermediate asymptotics to the solution to the equation of heat conduction with absorption,

$$\partial_t \theta = \kappa \partial_{xx}^2 \theta - \alpha \theta^n$$

where $n > 3$ is a constant, describing the decay of an amount of heat concentrated initially in a finite domain, is, as previously, the point-source self-similar solution of the first kind

$$\theta = \frac{Q}{\sqrt{\kappa t}} e^{-x^2/4\kappa t}.$$

For this distribution obviously the integral $M(t) = \int_{-\infty}^{\infty} \theta(x, t)dx = M_0$ is preserved in time. However, it is not preserved at the pre-self-similar stage, so that M_0 cannot be obtained by direct integration from the initial data. The numerical computation of a source-type initial-value problem can be recommended as an exercise.

and at the limit $\Pi_2 = \ell/(I\kappa t)^{1/3} \to 0$, which occurs when $t \to \infty$, a finite non-zero limit of the function $\Phi_{\max}(\Pi_2)$ does exist.

A researcher could easily be tempted to try to obtain in the same way, by dimensional analysis, a scaling law for the case of dome spreading with absorption. Indeed, in this case

$$H_{\max} = f(I, t, \kappa, \ell, c), \qquad \Pi_{\max} = \Phi_{\max}(\Pi_2, c) \qquad (4.15)$$

and it seems very natural to neglect Π_2 at large t and obtain the same law (4.13) with a constant depending on the absorption coefficient c. However, as follows from the results obtained in Chapter 3, at $t \to \infty$ there is a different scaling law, which cannot be obtained from dimensional analysis:

$$H_{\max} = \text{const} \left(\frac{I^{2(1-c)} \ell^{2c}}{\kappa t} \right)^{1/(3-2c)}. \qquad (4.16)$$

This scaling law for the decay in $H_{\max}$ is determined not just by the initial dome weight I but by the moment $I\ell^{c/(1-c)}$, and the power of ℓ in this moment cannot be obtained by dimensional analysis. Also, the constant in the scaling law can be obtained only by matching to the pre-self-similar solution.

The same situation exists with the spreading law for the water front. 'Naive' dimensional analysis gives the scaling law

$$x_f = \text{const}\,(I\kappa t)^{1/3} \qquad (4.17)$$

in both cases. However, a proper treatment (see Chapter 3) shows that the scaling law in the case of absorption is different:

$$x_f = \text{const}\,\left(I\ell^{c/(1-c)}\kappa t \right)^{(1-c)/(3-2c)}. \qquad (4.18)$$

Again, it is determined by κ, the time t and the same combined invariant $I\ell^{c/(1-c)}$.

The examples considered in Chapters 2 and 3 and discussed here are instructive. When we turn to the solution of a certain problem, and in particular to scaling laws, we do not know in advance to which type the solution belongs. The comparison of the cases of water-dome spreading with and without absorption considered above shows that the situation can be rather deceiving: from the point of view of whether it is possible to apply dimensional analysis these cases superficially do not differ from one another. But, as a matter of fact, as we have seen the scaling laws are quite different. Therefore it is necessary to keep in mind that it is a very strong hypothesis to assume the unimportance of certain governing parameters that would make the problem formulation non-idealized and its solution non-self-similar (in our case the initial dome width).

These governing parameters may be essential and yet self-similarity may nevertheless hold. Distinguishing between possible cases of self-similarity requires, in fact, a sufficiently deep mathematical investigation, which is unattainable in more complicated (especially nonlinear) problems. Therefore in obtaining self-similar solutions or scaling laws on the basis of dimensional analysis one should take care to verify, if only by means of numerical calculations or even experiments, that the solutions or scaling laws found actually reflect the required asymptotic behavior of the phenomenon under investigation. The situation is much more complicated if a mathematical formulation of the problem is lacking. In this case comparison with experimental data is of crucial importance. We will see this when considering the example of turbulent shear flow in Chapter 8.

4.3 A practical recipe for the application of similarity analysis

We have discussed above the fundamentals of dimensional analysis, similarity theory, scaling laws and self-similar phenomena. This discussion now allows us to provide a recipe for similarity analysis, i.e. for applying dimensional analysis and dealing with self-similarities and scaling laws. We emphasize that the basic difficulty always lies in finding an appropriate model, even a preliminary one. This is a matter of art, and no general recipe can be offered here. But when a researcher arrives at a particular model, and has the intention of working with this model, a certain general system of rules can be recommended.

Suppose that we are interested in a property a of some phenomenon (a may be a vector, i.e. there may be several such properties). We proceed in the following way.

1. *We specify a system of governing parameters $a_1, \ldots, a_k$ with independent dimensions and $b_1, \ldots, b_m$ with dependent dimensions such that a relation of the form*

$$a = f(a_1, \ldots, a_k, b_1, \ldots, b_m)$$

can be assumed to hold. If the model of the phenomenon has an explicit mathematical formulation then the independent variables and the constant parameters that appear in the equations, boundary conditions and initial conditions etc. are adopted as the governing parameters. If, however, there is no explicit mathematical formulation of the model then the governing parameters must be chosen on the basis of a qualitative model of the phenomenon, to be constructed by each investigator using his/her own experience and intuition as well as an analysis of previous studies.

2. *We choose an appropriate class of systems of units and determine the dimensions of the quantities under investigation, and of the governing parameters, in this class.* We then decide upon a system of governing parameters with independent dimensions: it is preferable to select those parameters whose importance to the phenomenon being studied is most firmly established.

3. *We express the dimensions of the quantities under investigation, and of the governing parameters with dependent dimensions, as products of powers of the dimensions of the governing parameters with independent dimensions.* We determine the similarity parameters and put the function under study into a dimensionless form, the *similarity* law

$$\Pi = \Phi(\Pi_1, \ldots, \Pi_m).$$

4. *We estimate the numerical values of the governing similarity parameters.* We select those that are large or small. In some cases, it turns out to be convenient at this stage to choose new similarity parameters that are products of powers of the similarity parameters obtained in the previous step: this sometimes makes it easier to perform these estimates.

5. *We try to formulate limiting similarity and scaling laws under the assumption of complete similarity in any large (or small) similarity parameters.* This means simply discarding these dimensionless governing parameters and the corresponding dimensional parameters. We compare the limiting similarity laws then obtained against the available experimental data and/or numerical calculations. If discrepancies are observed, we proceed as follows.

6. *We try to formulate limiting similarity laws under the assumption of incomplete similarity in the large (or small) similarity parameters.* This means that we assume a generalized homogeneity representation of the function $\Phi(\Pi_1, \ldots, \Pi_m)$ in terms of the small (or large) similarity parameters. Once again, we compare the similarity laws then obtained against the available experimental data, numerical calculations etc. If discrepancies are again observed, we can conclude that the phenomenon is not self-similar in the small (or large) similarity parameters. So, finally,

7. *We formulate general similarity laws and scaling laws using as few similarity parameters as possible.*

The use of this recipe will be demonstrated below in Chapters 7 and 8 with examples that are important in their own right, not merely as illustrations. It should be borne in mind that our attempt, in this section, to formalize the very informal procedure of finding ultimate similarity rules should be considered, naturally, as a general guide only. Here is another example of an attempt to formalize a very informal procedure. The Lawrence Berkeley National Laboratory, where the author is working, is situated in the East Bay Hills in Northern California. A few families of lions are also known to inhabit this place, therefore the Administration of the Laboratory issued a document giving rules for those who encounter a lion:

1. Do not hike alone.
2. Keep children close to you.
3. Do not approach the lion.
4. Do not run from the lion.
5. Do all you can to appear larger, and
6. **Fight back if attacked.**

Chapter 5
Scaling and transformation groups. Renormalization group

5.1 Dimensional analysis and transformation groups

We recall the definition of a transformation group. Suppose we have a set of transformations with k parameters,

$$x'_\nu = f_\nu(x_1, \ldots, x_n; A_1, \ldots, A_k), \qquad \nu = 1, \ldots, n, \tag{5.1}$$

where the f_ν are smooth functions of their arguments in a certain domain. We say that this set forms a k-parameter group of transformations if the following conditions are satisfied.

1. Among the transformations (5.1) there exists the identity transformation.
2. For each transformation of the set (5.1) there exists an inverse transformation that also belongs to the set (5.1).
3. For each pair of transformations of the set (5.1), i.e. a transformation **A** with parameters $A_1, \ldots, A_k$ and a transformation **B** with parameters $B_1, \ldots, B_k$, a transformation **C** with parameters $C_1, \ldots, C_k$, which also belongs to the set (5.1), exists and is uniquely determined such that successive realization of the transformations **A** and **B** is equivalent to the transformation **C**. The transformation **C** is called the *product* of the transformations **A** and **B**.

Dimensional analysis, which was considered in detail in Chapter 1, has a transparently group-theoretical nature. Group considerations can turn out to be useful also in those cases where dimensional analysis alone becomes insufficient to establish scaling laws and the self-similarity of a solution and to determine self-similar variables. A special place belongs here to the *renormalization group*, a concept now popular in theoretical physics.

Dimensional analysis is based on the Π-theorem (see Chapter 1). This theorem allows one to express a dimensional, generally speaking, function of

94

$n = k + m$ dimensional governing parameters, i.e. the physically meaningful relationship

$$a = f(a_1, \ldots, a_k, b_1, \ldots, b_m) \qquad (5.2)$$

where $a_1, \ldots, a_k$ are the governing parameters with independent dimensions, as a dimensionless function of m dimensionless parameters

$$\Pi = \Phi(\Pi_1, \ldots, \Pi_m),$$

where

$$\Pi = \frac{a}{a_1^p \cdots a_k^r}, \qquad \Pi_1 = \frac{b_1}{a_1^{p_1} \cdots a_k^{r_1}}, \qquad \ldots, \qquad \Pi_m = \frac{b_m}{a_1^{p_m} \cdots a_k^{r_m}}.$$

This means that the function f in (5.2) possesses the property of generalized homogeneity:

$$f(a_1, \ldots, a_k, b_1, \ldots, b_m) = a_1^p \cdots a_k^r \, \Phi \left(\frac{b_1}{a_1^{p_1} \cdots a_k^{r_1}}, \ldots, \frac{b_m}{a_1^{p_m} \cdots a_k^{r_m}} \right).$$

We note now that, for any positive numbers $A_1, \ldots, A_k$, the scaling transformation of the governing parameters with independent dimensions

$$a_1' = A_1 a_1, \qquad a_2' = A_2 a_2, \qquad \ldots, \qquad a_k' = A_k a_k \qquad (5.3)$$

can be obtained by changing from the original system of units to some other system belonging to the same class. At the same time the values of the remaining parameters $a, b_1, \ldots, b_m$ vary in accordance with their dimensions:

$$a' = A_1^p \cdots A_k^r a,$$
$$b_1' = A_1^{p_1} \cdots A_k^{r_1} b_1,$$
$$\vdots$$
$$b_m' = A_1^{p_m} \cdots A_k^{r_m} b_m. \qquad (5.4)$$

Direct verification shows easily that the transformations (5.3), (5.4) form a k-parameter group. Indeed, if $A_1 = A_2 = \cdots = A_k = 1$ then the transformation (5.3), (5.4) becomes an identity transformation. For each transformation **A** in the set (5.3), (5.4) there exists an inverse transformation **B** with parameter values

$$B_1 = \frac{1}{A_1}, \qquad B_2 = \frac{1}{A_2}, \qquad \ldots, \qquad B_k = \frac{1}{A_k}$$

which also belongs to the set (5.3), (5.4) and such that the successive realization of transformations **A** and **B** returns the variables $x_1 \cdots x_n$ to their original values. For each pair **A**, **B** of transformations (5.3), (5.4), with parameter values $A_1, \ldots, A_k$ and $B_1, \ldots, B_k$, there exists one and only one transformation **C**, with parameter values $C_1 = A_1 B_1, C_2 = A_2 B_2, \ldots, C_k = A_k B_k$, also belonging to the class (5.3), (5.4) and such that the successive realization of transformations **A** and **B** is equivalent to the transformation **C**.

The quantities Π, Π_1, ..., Π_m remain unchanged for all transformations of the group (5.3), (5.4), i.e. they are *invariants* of this group. Thus, the Π-theorem is a simple consequence of the covariance principle: relations with a physical meaning among dimensional quantities of the form (5.2) can be represented in a form invariant with respect to the group of similarity transformations of the governing parameters with independent dimensions (5.3), (5.4), each transformation corresponding to a transition to a different system of units (within a given class). The number of independent invariants of the group, i.e. the number of invariants which cannot be obtained as a product of power of the other ones, is less than the total number of governing parameters by the number k of parameters of the group.

The invariance of the formulation, and hence the solution, of any physically meaningful problem with respect to the group of transformations (5.3), (5.4) is thus necessary according to the general physical covariance principle. It can turn out, however, that there exists a richer group with respect to which the formulation of the special problem considered is invariant. Then the number of arguments of the function Φ in the universal (invariant) relation obtained after applying the Π-theorem in its own right should be reducible by the number of parameters of the supplementary group. Here the solution can turn out to be self-similar, and the self-similar variables can be determined as a result of using the invariance with respect to the supplementary group, although this self-similarity is not implied by dimensional analysis (which exploits invariance with respect to the group of similarity transformations of the governing parameters with independent dimensions). We consider below an instructive example that will clarify this idea.

5.2 Problem: the boundary layer on a flat plate in a uniform flow

The problem of steady viscous incompressible flow past a semi-infinite flat plate placed along a uniform stream (Figure 5.1) leads to a system of Navier–Stokes equations and the equation of continuity (see Batchelor 1967; Germain 1986; Landau and Lifshitz 1987):

$$u\partial_x u + v\partial_y u = -\frac{1}{\rho}\,\partial_x p + v(\partial_{xx}^2 u + \partial_{yy}^2 u),$$

$$u\partial_x v + v\partial_y v = -\frac{1}{\rho}\,\partial_y p + v(\partial_{xx}^2 v + \partial_{yy}^2 v),$$

$$\partial_x u + \partial_y v = 0.$$

(5.5)

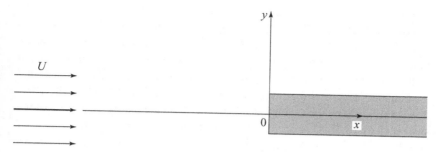

Figure 5.1. Viscous flow past a thin semi-infinite plate.

Here x and y are the longitudinal and transverse Cartesian coordinates, $u(x, y)$ and $v(x, y)$ are the corresponding velocity components, p is the pressure, v is the kinematic viscosity coefficient and ρ is the density of the fluid.

The boundary conditions for the problem under consideration can be represented in the form

$$u(x, 0) = v(x, 0) = 0, \qquad x \geq 0,$$
$$u(x, y) \to U, \qquad v(x, y) \to 0 \qquad \text{for } y^2 \to \infty \text{ and arbitrary } x$$
$$\text{and for } x \to -\infty \text{ and arbitrary } y.$$

Here U is the constant speed of the uniform exterior flow; the origin of coordinates $x = 0$, $y = 0$ corresponds to the tip of the plate. Up to now no single problem of viscous flow past a body has been solved analytically; the problem of the flow past a semi-infinite plate presented above, in spite of its seeming simplicity, does not constitute an exception.

At the beginning of the last century Prandtl (1904) proposed the idea of the *boundary layer*, which revolutionized fluid mechanics as a whole and, in particular, led to an asymptotic approximate analytic solution of the problem of viscous flow past a plate. This solution was obtained by Prandtl's student Blasius (1908); Blasius' analysis was modified by Toepfer (1912). The basic model of Prandtl in application to this problem is that at large Reynolds number the effects of viscosity are concentrated in a thin layer surrounding the plate only. Prandtl's hypothesis and certain qualitative considerations allowed a reduction of the model to a simplified one (see Batchelor 1967; Schlichting 1968; Germain 1986; Landau and Lifshitz 1987), the system of equations

$$u \partial_x u + v \partial_y u = v \partial_{yy}^2 u, \qquad \partial_x u + \partial_y v = 0 \tag{5.6}$$

under boundary conditions at $x \geq 0$, $y \geq 0$

$$u(0, y) = U, \qquad u(x, \infty) = U, \qquad u(x, 0) = v(x, 0) = 0. \tag{5.7}$$

Two comments: first, no one has been able, up to now, to give a rigorous mathematical derivation of the system (5.6), (5.7) from the Navier–Stokes equations at large Reynolds numbers without additional assumptions – this system remains a result of Prandtl's intuition. The second comment concerns the second of the boundary conditions (5.7), which seems paradoxical: it is claimed that the boundary layer is thin yet the condition is taken at infinity. In fact, this paradox is explained by the asymptotic character of the qualitative derivation of the system (5.6), (5.7). This derivation is based on a 'stretching' of the system of coordinates, an asymptotic analysis of the problem in the stretched coordinates and a subsequent return to the original coordinates. This asymptotic procedure is illuminated by an original example proposed by Friedrichs (1966).

We apply to the problem in the boundary-layer approximation (5.6), (5.7) the standard procedure of dimensional analysis. The governing parameters are ν, x, U and y, so that

$$u = f_u(\nu, x, U, y), \qquad v = f_v(\nu, x, U, y). \qquad (5.8)$$

The dimensions of the involved quantities are

$$[u] = [v] = [U] = \frac{L}{T}, \qquad [x] = [y] = L, \qquad [\nu] = \frac{L^2}{T} \qquad (5.9)$$

so that, according to the standard procedure of dimensional analysis,

$$\Pi_u = \frac{u}{U} = \Phi_u(\Pi_1, \Pi_2), \qquad \Pi_v = \frac{v}{U} = \Phi_v(\Pi_1, \Pi_2). \qquad (5.10)$$

Here

$$\Pi_1 = \xi = \frac{Ux}{\nu}, \qquad \Pi_2 = \eta = \frac{Uy}{\nu}. \qquad (5.11)$$

In the new variables the relations (5.6), (5.7) are reduced to the form

$$\Phi_u \partial_\xi \Phi_u + \Phi_v \partial_\eta \Phi_u = \partial_{\eta\eta}^2 \Phi_u, \qquad \partial_\xi \Phi_u + \partial_\eta \Phi_v = 0$$
$$\Phi_u(0, \eta) = \Phi_u(\xi, \infty) = 1, \qquad \Phi_u(\xi, 0) = \Phi_v(\xi, 0) = 0 \quad (5.12)$$

We see that the direct application of dimensional analysis does not give any simplification of the problem. In fact, the only distinction between the system (5.6), (5.7) and the system (5.12) is that in the latter the constants ν and U are equal to unity: a purely cosmetic transformation.

It is instructive, however, that the system (5.12) is invariant with respect to an additional transformation group. Indeed, let $\Phi_u(\xi, \eta)$, $\Phi_v(\xi, \eta)$ be a solution of the system (5.12) which exists and is unique. Let us consider a one-parameter transformation group:

$$\xi' = \alpha^2 \xi, \qquad \eta' = \alpha \eta;$$
$$\Phi_u'(\xi', \eta') = \Phi_u(\xi, \eta), \qquad \Phi_v'(\xi', \eta') = \alpha^{-1} \Phi_v(\xi, \eta), \qquad (5.13)$$

where $\alpha > 0$ is the parameter. It is easy to verify by direct substitution that the set of transformations (5.13) is a group: $\alpha = 1$ gives the identical transformation, $\beta = \alpha^{-1}$ gives the transformation inverse to α and $\gamma = \alpha\beta$ gives the product of the transformations with parameters α and β. Substituting (5.13) into (5.12), we obtain for arbitrary positive α the same problem as (5.12) but in the variables $\xi', \eta', \Phi_u', \Phi_v'$. In view of the uniqueness requirement, the solution Φ_u', Φ_v' should also be unique, so that

$$\Phi_u(\xi, \eta) = \Phi_u'(\xi', \eta') = \Phi_u(\alpha^2 \xi, \alpha \eta),$$
$$\Phi_v(\xi, \eta) = \alpha \Phi_v'(\xi', \eta') = \alpha \Phi_v(\alpha^2 \xi, \alpha \eta). \qquad (5.14)$$

Furthermore, *after establishing* the relations (5.14) the value of the parameter α can be taken as equal to an arbitrary positive number, in particular

$$\alpha = \frac{1}{\sqrt{\xi}}.$$

Substituting this relation into (5.14), we obtain that *the determination of the functions Φ_u, Φ_v of two variables is reduced to the determination of functions of a single variable*

$$\Phi_u(\xi, \eta) = \Phi_u\left(1, \frac{\eta}{\sqrt{\xi}}\right) = f_u\left(\frac{\eta}{\sqrt{\xi}}\right) = f_u\left(\frac{y}{\sqrt{\nu x/U}}\right)$$

$$\Phi_v(\xi, \eta) = \frac{1}{\sqrt{\xi}}\Phi_v\left(1, \frac{\eta}{\sqrt{\xi}}\right) = \frac{1}{\sqrt{\xi}} f_v\left(\frac{\eta}{\sqrt{\xi}}\right) = \sqrt{\frac{\nu}{Ux}} f_v\left(\frac{y}{\sqrt{\nu x/U}}\right).$$

$$(5.15)$$

As we see, the solution is self-similar. Thus the self-similarity of the solution to the boundary-layer problem (5.6), (5.7) is established and the expressions for the self-similar variables are obtained. However, it has been achieved as a result of the application of not only dimensional analysis but also the invariance of the problem with respect to an additional transformation group (5.13).

Introducing a new function

$$\varphi(\zeta) = \int_0^\zeta f_u(\zeta)\, d\zeta, \qquad \zeta = \frac{\eta}{\sqrt{\xi}} = \frac{y}{\sqrt{\nu x/U}},$$

we obtain from (5.6), (5.7) and the definition of the function $\varphi(\zeta)$ the relations

$$f_v = \frac{1}{2}(\zeta\varphi' - \varphi), \tag{5.16}$$

$$\varphi\varphi'' + 2\varphi''' = 0, \qquad \varphi(0) = \varphi'(0) = 0, \quad \varphi'(\infty) = 1; \tag{5.17}$$

here a prime indicates differentiation. The relationships (5.17) present a *boundary-value problem* for the ordinary equation $\varphi\varphi'' + 2\varphi''' = 0$, with boundary-value data at both $\zeta = 0$ and $\zeta = \infty$. This is inconvenient, and here also a simple group-theoretical consideration is helpful. Indeed, let us consider the family of solutions to the equation $\varphi\varphi'' + 2\varphi''' = 0$ satisfying the *two boundary conditions* at $\zeta = 0$, $\varphi(0) = \varphi'(0) = 0$. It is easy to check that this family is invariant with respect to the transformation group:

$$\varphi_1(\zeta_1) = \alpha^{-1}\varphi(\zeta), \qquad \zeta_1 = \alpha\zeta, \tag{5.18}$$

so that if $\varphi(\zeta)$ is a solution to the equation $\varphi\varphi'' + 2\varphi''' = 0$ satisfying the first two boundary conditions in (5.17), then for any positive α the function $\alpha\varphi(\alpha\zeta)$ also satisfies the equation and these two boundary conditions.

Now consider the solution $\varphi_0(\zeta)$ to the Cauchy (not the boundary-value) problem for which the third boundary condition in (5.17), the condition at infinity, *is replaced by a condition at zero*, $\varphi_0''(0) = 1$. For the solution $\varphi_0(\zeta)$, which is easy to calculate numerically, the value of the derivative at infinity, $\varphi_0'(\infty)$, is 2.086. Therefore the solution $\varphi(\zeta) = \alpha\varphi_0(\alpha\zeta)$, where $\alpha = 1/\sqrt{2.086} = 0.6925$, satisfies all the conditions of problem (5.17).

For the drag F on a section of unit width and length l of the flat plate in a uniform stream of velocity U we obtain from the previous relations, using the results of numerical calculation of the function φ,

$$F = 2\int_0^l (\sigma_{xy})_{y=0}\, dx = 2U\sqrt{\frac{U}{\nu}}\, \rho\nu \int_0^l f_u'(0)\,\frac{dx}{\sqrt{x}}$$

$$= 4\sqrt{\frac{U^3 l}{\nu}}\, \rho\nu\varphi''(0) = 4\alpha^3 \rho\sqrt{U^3 l\nu}$$

$$= 1.328\rho\sqrt{U^3 l\nu}.$$

Here $(\sigma_{xy})_{y=0} = \rho\nu(\partial_y u)_{y=0}$ is the shear stress on the plate.

Introducing the dimensionless parameter $\Pi = F/(\rho U^2 l)$ corresponding to the drag F, we get

$$\Pi = \Phi(Re) = \frac{1.328}{\sqrt{Re}}, \qquad Re = \frac{Ul}{\nu}.$$

We note in passing that one can also see that at this well-known relation reveals incomplete similarity in the Reynolds number. In fact, the drag F is determined by the following quantities: the length l of the plate, the viscosity ν and density ρ of the fluid and the velocity U of the stream. Application of the standard procedure of dimensional analysis gives

$$\Pi = \Phi(Re).$$

For the high Reynolds numbers characteristic of the boundary layer there is no complete similarity with respect to Reynolds number, since there does not exist a non-zero limit of the function $\Phi = 1.328 Re^{-1/2}$ as $Re \to \infty$. Hence the relations

$$\Pi = \text{const}, \qquad F = \text{const } \rho U^2 l$$

that would have to hold in the case of complete similarity in the Reynolds number cannot be expected to be true, no matter how high the Reynolds number. Nevertheless, one has the relation

$$\Pi^* = \frac{F}{\rho \sqrt{U^3 l \nu}} = \text{const} = 1.328,$$

corresponding to incomplete self-similarity: the parameter Π^* cannot be obtained from standard dimensional analysis and contains the dimensional parameter ν whose explicit introduction into the problem violates self-similarity.

The example of boundary-layer flow past a flat plate which we have just considered is instructive also in the following aspect: the application of a more general group of transformations can here be given the form of *generalized dimensional analysis*, and this device turns out to be useful in many other special cases (but, it should be emphasized, not always).

Namely, we shall use different units to measure length in the x-direction and length in the y-direction. So, we introduce two different units of length, l_x and l_y, and consider x and y as having different dimensions L_x and L_y. Let us use this in the boundary-layer problem (5.6), (5.7). In this case all terms entering the boundary-layer equations and boundary conditions of the problem have identical dimensions if we take $[u] = [U] = L_x/T$, $[v] = L_y^2/T$, $[\nu] = L_y/T$, $[x] = L_x$ and $[y] = L_y$. Thus, among the governing parameters ν, x, U and y not two but three have independent dimensions and the single independent dimensionless similarity parameter will be

$$\Pi_1' = \zeta = \frac{y}{\sqrt{\nu x / U}}, \tag{5.19}$$

whence follows immediately the self-similarity of the solution

$$u = U f_u(\zeta), \qquad v = \sqrt{\nu U / x} \, f_v(\zeta). \tag{5.20}$$

It is instructive that using such independent units for longitudinal and transverse lengths is impossible for the full Navier–Stokes equations (5.5). In these equations the terms $\nu\partial^2_{yy}u$ and $\nu\partial^2_{yy}v$ appear in sum with the terms $\nu\partial^2_{xx}u$ and $\nu\partial^2_{xx}v$, so that if we measure x and y in different units these terms will have different dimensions, and this is impossible for equations having physical meaning. Consequently, the full Navier–Stokes equations, unlike the boundary layer equations, are not invariant with respect to the transformation group (5.13).

A natural question arises: is there an algorithm for seeking a maximally broad group of transformations with respect to which a given system of differential equations is invariant? Such an algorithm does exist. The basic ideas here belong to the Norwegian mathematician of the nineteenth century Sophus Lie. In recent times a series of general results and applications to particular systems of equations encountered in applied mathematics have been obtained; we refer the reader to the valuable books by Birkhof (1960), Bluman and Cole (1974) and Olver (1993).

5.3 The renormalization group and incomplete similarity

5.3.1 The renormalization group and intermediate asymptotics

Among the groups additional to the compulsory group of scaling transformations of quantities with independent dimensions that lead to scaling laws and self-similarity, a special and very important place belongs to the renormalization group. The renormalization group approach, following the ideas of Stückelberg and Peterman (1953), Gell-Mann and Low (1954), Bogolyubov and Shirkov (1955, 1959), Kadanoff (1966), Kadanoff et al. (1967), Patashinsky and Pokrovsky (1966) and Wilson (1971), has found extensive applications in modern theoretical physics. N. Goldenfeld, O. Martin and Y. Oono demonstrated a deep relation between the renormalization group method as traditionally used by physicists and the intermediate-asymptotics approach, developed independently and presented in this book. They did this by using the renormalization group method, in the form in which it is usually applied by physicists to solve some typical problems whose solution had been obtained previously by the method of intermediate asymptotics. Vice versa, they solved by the method of intermediate asymptotics some classical problems in statistical physics solved earlier by the renormalization group approach (Goldenfeld 1989; Goldenfeld, Martin and Oono 1989, 1991; Goldenfeld et al. 1990; Goldenfeld and Oono 1991; Chen, Goldenfeld and Oono 1991; Chen and Goldenfeld 1992; the book Goldenfeld 1992 is especially recommended).

We recall, see Chapters 1 and 4 and section 5.1 of this chapter, that any physically significant relation among dimensional (generally speaking) parameters

$$a = f(a_1, \ldots a_k, b_1, \ldots, b_m)$$

can be represented in the form of a relation between normalized dimensionless parameters Π, Π_i, $i = 1, \ldots, m$:

$$\Pi = \Phi(\Pi_1, \ldots, \Pi_m).$$

This is due to the compulsory invariance of physically significant relations with respect to the transformation group (5.3), (5.4) corresponding to a transition from the original system of units of measurement to an arbitrary system of units belonging to the same class of systems of units, i.e. having basic units of the same physical nature but different magnitude.

This means, we repeat, that every function f which enters a physically significant relation possesses the property of generalized homogeneity:

$$f(a_1, \ldots, a_k, b_1, \ldots, b_m) = a_1^p \cdots a_k^r \Phi\left(\frac{b_1}{a_1^{p_1} \cdots a_k^{r_1}}, \ldots, \frac{b_m}{a_1^{p_m} \cdots a_k^{r_m}}\right).$$

In the general case of incomplete similarity the function Φ possesses at large or small values of the dimensionless parameters $\Pi_{\ell+1}, \ldots, \Pi_m$ the same property of generalized homogeneity in its own renormalized dimensionless arguments:

$$
\begin{aligned}
&\Phi(\Pi_1, \ldots, \Pi_\ell, \Pi_{\ell+1}, \ldots, \Pi_m) \\
&= \Pi_{\ell+1}^{\alpha_{\ell+1}} \cdots \Pi_m^{\alpha_m} \Phi_1\left(\frac{\Pi_1}{\Pi_{\ell+1}^{\beta_1} \cdots \Pi_m^{\delta_1}}, \ldots, \frac{\Pi_\ell}{\Pi_{\ell+1}^{\beta_\ell} \cdots \Pi_m^{\delta_\ell}}\right) \quad (5.21)
\end{aligned}
$$

where the powers $\alpha_{\ell+1}, \ldots, \delta_\ell$ are certain constants, which cannot be obtained by dimensional analysis even in principle.

The property of incomplete similarity also has a group-theoretical nature. It means that in addition to the compulsory group of transformations (5.3), (5.4) the problem at large or small values of the dimensionless parameters $\Pi_{\ell+1}, \ldots, \Pi_m$ has the property of invariance with respect to the set of transformations

$$
\begin{aligned}
a_1' &= a_1, & a_2' &= a_2, & \ldots, & & a_k' &= a_k, \\
b_1' &= B_{\ell+1}^{\beta_1} \cdots B_m^{\delta_1} b_1, & & & \ldots, & & b_\ell' &= B_{\ell+1}^{\beta_\ell} \cdots B_m^{\delta_\ell} b_\ell, \\
b_{\ell+1}' &= B_{\ell+1} b_{\ell+1}, & & & \ldots, & & b_m' &= B_m b_m, \\
a' &= B_{\ell+1}^{\alpha_{\ell+1}} \cdots B_m^{\alpha_m} a. & & & & & & & (5.22)
\end{aligned}
$$

Here the parameters $B_{\ell+1}, \ldots, B_m$ are certain positive numbers. Naturally, the values of these parameters should not be too large or too small, otherwise the applicability of the asymptotics (5.21) will be violated. The set (5.22) has the properties of a transformation group with parameters $B_{\ell+1}, \ldots, B_m$. Indeed, if all the $B_{\ell+1}, \ldots, B_m$ are equal to unity then the transformation (5.22) is an identity transformation. For every transformation in the set (5.22) there exists an inverse transformation with parameters $B_{\ell+1}^{-1}, \ldots, B_m^{-1}$, also belonging to this set. Finally, the product of two transformations with parameters $B_{\ell+1}^{(1)}, \ldots, B_m^{(1)}$ and $B_{\ell+1}^{(2)}, \ldots, B_m^{(2)}$, which has parameters $B_{\ell+1} = B_{\ell+1}^{(1)} B_{\ell+1}^{(2)}, \ldots, B_m = B_m^{(1)} B_m^{(2)}$, also exists in the set (5.22) and is uniquely determined. We will identify the group (5.22) with the renormalization group and so establish a link between this concept and the concepts of intermediate asymptotics and incomplete similarity considered earlier in this book.

More precisely, we will prove that the statement of the asymptotic invariance to the renormalization group (5.22) of the basic relation obtained after the application of dimensional analysis,

$$\Pi = \Phi(\Pi_1, \ldots, \Pi_\ell, \Pi_{\ell+1}, \ldots, \Pi_m), \qquad (5.23)$$

is equivalent to the statement of incomplete similarity.

Indeed, assume that there is incomplete similarity in the parameters $\Pi_{\ell+1}, \ldots, \Pi_m$ at small values, for definiteness sake, of these parameters, i.e. that the relation (5.21) holds for the function Φ. Let us perform the transformations (5.22). We form the quantities

$$\Pi_1' = \frac{b_1'}{a_1'^{p_1} \cdots a_k'^{r_1}} = B_{\ell+1}^{\beta_1} \cdots B_m^{\delta_1} \frac{b_1}{a_1^{p_1} \cdots a_k^{r_1}} = B_{\ell+1}^{\beta_1} \cdots B_m^{\delta_1} \Pi_1;$$

$$\vdots$$

$$\Pi_\ell' = \frac{b_\ell'}{a_1'^{p_\ell} \cdots a_k'^{r_\ell}} = B_{\ell+1}^{\beta_\ell} \cdots B_m^{\delta_\ell} \frac{b_\ell}{a_1^{p_\ell} \cdots a_k^{r_\ell}} = B_{\ell+1}^{\beta_\ell} \cdots B_m^{\delta_\ell} \Pi_\ell,$$

$$\Pi_{\ell+1}' = \frac{b_{\ell+1}'}{a_1'^{p_{\ell+1}} \cdots a_k'^{r_{\ell+1}}} = B_{\ell+1} \frac{b_{\ell+1}}{a_1^{p_{\ell+1}} \cdots a_k^{r_{\ell+1}}} = B_{\ell+1} \Pi_{\ell+1};$$

$$\vdots$$

$$\Pi_m' = \frac{b_m'}{a_1'^{p_m} \cdots a_k'^{r_m}} = B_m \frac{b_m}{a_1^{p_m} \cdots a_k^{r_m}} = B_m \Pi_m,$$

$$\Pi' = B_{\ell+1}^{\alpha_{\ell+1}} \cdots B_m^{\alpha_m} \frac{a}{a_1^p \cdots a_k^r} = \frac{a'}{a_1'^p \cdots a_k'^r} = B_{\ell+1}^{\alpha_{\ell+1}} \cdots B_m^{\alpha_m} \Pi.$$

Clearly, for every $i = 1, \ldots, \ell$ we have by construction

$$\frac{\Pi'_i}{\Pi'^{\beta_i}_{\ell+1} \cdots \Pi'^{\delta_i}_m} = \frac{\Pi_i}{\Pi^{\beta_i}_{\ell+1} \cdots \Pi^{\delta_i}_m}$$

and also

$$\frac{\Pi'}{\Pi'^{\alpha_{\ell+1}}_{\ell+1} \cdots \Pi'^{\alpha_m}_m} = \frac{\Pi}{\Pi^{\alpha_{\ell+1}}_{\ell+1} \cdots \Pi^{\alpha_m}_m}.$$

We obtain using (5.21),

$$\begin{aligned}
\Pi' &= B^{\alpha_{\ell+1}}_{\ell+1} \cdots B^{\alpha_m}_m \Pi = B^{\alpha_{\ell+1}}_{\ell+1} \cdots B^{\alpha_m}_m \Phi(\Pi_1, \ldots, \Pi_\ell, \Pi_{\ell+1}, \ldots, \Pi_m) \\
&= B^{\alpha_{\ell+1}}_{\ell+1} \cdots B^{\alpha_m}_m \Pi^{\alpha_{\ell+1}}_{\ell+1} \cdots \Pi^{\alpha_m}_m \Phi_1\left(\frac{\Pi_1}{\Pi^{\beta_1}_{\ell+1} \cdots \Pi^{\delta_1}_m}, \ldots, \frac{\Pi_\ell}{\Pi^{\beta_\ell}_{\ell+1} \cdots \Pi^{\delta_\ell}_m}\right) \\
&= \Pi'^{\alpha_{\ell+1}}_{\ell+1} \cdots \Pi'^{\alpha_m}_m \Phi_1\left(\frac{\Pi'_1}{\Pi'^{\beta_1}_{\ell+1} \cdots \Pi'^{\delta_1}_m}, \ldots, \frac{\Pi'_\ell}{\Pi'^{\beta_\ell}_{\ell+1} \cdots \Pi'^{\delta_\ell}_m}\right) \\
&= \Phi(\Pi'_1, \ldots \Pi'_\ell, \Pi'_{\ell+1}, \ldots, \Pi'_m).
\end{aligned}$$

Thus, from incomplete similarity, (5.21), follows the invariance of the basic relation (5.23) with respect to the renormalization group (5.22). And now, vice versa, assume that there is invariance of the basic relation (5.23) with respect to the group (5.22). This means that for every $B_{\ell+1}, \ldots, B_m$ the relation (5.23) preserves its form after the transformation (5.22). Without loss of generality we can rewrite (5.23) in the form

$$\frac{\Pi'}{\Pi'^{\alpha_{\ell+1}}_{\ell+1} \cdots \Pi'^{\alpha_m}_m} = \Psi\left(\frac{\Pi'_1}{\Pi'^{\beta_1}_{\ell+1} \cdots \Pi'^{\delta_1}_m}, \ldots, \frac{\Pi'_\ell}{\Pi'^{\beta_\ell}_{\ell+1} \cdots \Pi'^{\delta_\ell}_m}, \Pi'_{\ell+1}, \ldots, \Pi'_m\right).$$

Returning to the previous variables we obtain

$$\begin{aligned}
\frac{\Pi'}{\Pi'^{\alpha_{\ell+1}}_{\ell+1} \cdots \Pi'^{\alpha_m}_m} &= \frac{\Pi}{\Pi^{\alpha_{\ell+1}}_{\ell+1} \cdots \Pi^{\alpha_m}_m} \\
&= \frac{\Phi(\Pi_1, \ldots, \Pi_\ell, \Pi_{\ell+1}, \ldots, \Pi_m)}{\Pi^{\alpha_{\ell+1}}_{\ell+1} \cdots \Pi^{\alpha}_m} \\
&= \Psi\left(\frac{\Pi_1}{\Pi^{\beta_1}_{\ell+1} \cdots \Pi^{\delta_1}_m}, \ldots, \frac{\Pi_\ell}{\Pi^{\beta_\ell}_{\ell+1} \cdots \Pi^{\delta_\ell}_m}, B_{\ell+1}\Pi_{\ell+1}, \ldots, B_m\Pi_m\right).
\end{aligned}$$

From this relation it follows (compare the proof of the basic theorem of dimensional analysis in Chapter 1) that the function Ψ does not depend on the arguments $\Pi'_{\ell+1}, \ldots, \Pi'_m$. Indeed, let us fix all parameters B_i, $i = \ell+1, \ldots, m$, except for one, say B_j, and vary B_j arbitrarily. The result will not depend on B_j. Therefore

$$\Psi = \Phi_1 \left(\frac{\Pi_1}{\Pi_{\ell+1}^{\beta_1} \cdots \Pi_m^{\delta_1}}, \ldots, \frac{\Pi_\ell}{\Pi_{\ell+1}^{\beta_\ell} \cdots \Pi_m^{\delta_\ell}}, \right).$$

Thus the function Φ has the property of generalized homogeneity (5.21) and we have a case of incomplete similarity. We have proved the equivalence of incomplete similarity and invariance with respect to the renormalization group.

5.3.2 The perturbation expansion

The basic relation (5.2) in which we are interested can be written in a dimensionless form as

$$\Pi = \Phi(\Pi_1, \ldots, \Pi_m, c).$$

Here we have added an additional constant dimensionless parameter c on which the phenomenon is also assumed to depend. Its use will become clear shortly. Again let the parameters $\Pi_{\ell+1}, \ldots \Pi_m$ be small for the sake of definiteness and assume that generally speaking, incomplete similarity holds, so that

$$\Pi = \Phi(\Pi_1, \ldots, \Pi_\ell, \Pi_{\ell+1}, \ldots, \Pi_m, c)$$
$$= \Pi_{\ell+1}^{\alpha_{\ell+1}} \cdots \Pi_m^{\alpha_m} \Phi_1 \left(\frac{\Pi_1}{\Pi_{\ell+1}^{\beta_1} \cdots \Pi_m^{\delta_1}}, \ldots, \frac{\Pi_\ell}{\Pi_{\ell+1}^{\beta_\ell} \cdots \Pi_m^{\delta_\ell}}, c \right);$$

it follows that at least one of the powers $\alpha_{\ell+1}, \ldots, \delta_\ell$ is different from zero. Generally speaking, $\alpha_{\ell+1}, \ldots, \delta_\ell$ depend on the parameter c. Let us assume further that all the powers $\alpha_{\ell+1}, \ldots, \delta_\ell$ vanish at $c = 0$, i.e. that at $c = 0$ we have a case of complete similarity . Then, for sufficiently small $\Pi_{\ell+1}, \ldots, \Pi_m$, the function $\Phi(\Pi_1, \ldots, \Pi_\ell, \Pi_{\ell+1}, \ldots, \Pi_m)$ can be replaced by its finite non-zero limit $\Phi(\Pi_1, \ldots, \Pi_\ell, 0, \ldots, 0)$, so that the dimensional parameters $b_{\ell+1}, \ldots, b_m$ disappear from consideration. We can say, therefore, that at sufficiently small values for $\Pi_{\ell+1}, \ldots, \Pi_m$ the phenomenon is asymptotically invariant to the transformation group

$$a' = a, \qquad a_1' = a_1, \qquad \ldots, \qquad a_k' = a_k,$$
$$b_1' = b_1, \qquad \ldots, \qquad b_\ell' = b_\ell; \qquad b_{\ell+1}' = B_{\ell+1} b_{\ell+1}, \qquad \ldots, \qquad b_m' = B_m b_m,$$
$$(5.24)$$

where $B_{\ell+1}, \ldots, B_m$ are the group parameters. In the case of incomplete similarity the problem is asymptotically invariant with respect to a more complicated renormalization group, (5.22).

The next step and, we emphasize, an independent one, is to obtain the parameters $\alpha_{\ell+1}, \ldots, \delta_\ell$ by a perturbation expansion, using some quantitative relations concerning the phenomenon, in particular, the non-integrable conservation laws. The latter point is crucial: if no further information concerning the phenomenon under consideration is available then the parameters $\alpha_{\ell+1}, \ldots, \delta_\ell$ entering the renormalization group (5.22) and the incomplete similarity relation (5.21) cannot be determined.

As an example we will consider a perturbation expansion for the problem of groundwater dome spreading with absorption considered in Chapter 3.

From the basic equation for the water head (3.5) the non-integrable (generally speaking) conservation law (3.8) was obtained:

$$\frac{d}{dt} \int_{-x_f}^{x_f} H(x, t)\, dx = -2\kappa c \int_{-x_f}^{x_f} (\partial_x H)^2 dx.$$

The limiting self-similar solution was represented in the form (3.21):

$$H = \frac{\xi_f^2 \left(I \ell^{(1-3\mu)/\mu}\right)^{2\mu} \mu}{(\kappa t)^{1-2\mu}}\, f(\zeta, c),$$

$$\zeta = \frac{x}{x_f}, \qquad x_f = \xi_f \left(I \ell^{(1-3\mu)/\mu} \kappa t\right)^{\mu}.$$

$$(5.25)$$

Therefore

$$\int_{-x_f}^{x_f} H(x, t)\, dx = \frac{\xi_f^3 \left(I \ell^{(1-3\mu)/\mu}\right)^{3\mu} \mu}{(\kappa t)^{1-3\mu}} \left(\int_{-1}^{1} f(\zeta, c)\, d\zeta\right),$$

$$\int_{-x_f}^{x_f} (\partial_x H)^2 dx = \frac{\xi_f^3 \left(I \ell^{(1-3\mu)/\mu}\right)^{3\mu} \mu^2}{(\kappa t)^{2-3\mu}} \left(\int_{-1}^{1} [f'(\zeta, c)]^2 d\zeta\right).$$

At $c = 0$, $\mu = 1/3$; therefore the value of $1 - 3\mu$ is small at small values of c. To the accuracy of the leading terms we obtain

$$\frac{d}{dt} \int_{-x_f}^{x_f} H(x, t)\, dx = -\frac{1 - 3\mu}{3(\kappa t)^{2-3\mu}} \kappa \xi_f^3 I \int_{-1}^{1} f(\zeta, 0)\, d\zeta,$$

$$\int_{-x_f}^{x_f} (\partial_x H)^2 dx = \frac{\xi_f^3 I}{9(\kappa t)^{2-3\mu}} \int_{-1}^{1} [f'(\zeta, 0)]^2 d\zeta.$$

Now we need to use the relations

$$f(\zeta, 0) = \frac{1}{4}(1 - \zeta^2), \qquad f'(\zeta, 0) = -\frac{1}{2}\zeta,$$

$$\int_{-1}^{1} f \, d\zeta = \frac{1}{3}, \qquad \int_{-1}^{1} (f')^2 d\zeta = \frac{1}{6}.$$

We substitute these relations into the non-integrable conservation law (3.8) and obtain

$$-\frac{1}{9}(1 - 3\mu) = -\frac{1}{27}c, \qquad \text{so that} \qquad \mu = \frac{1}{3}\left(1 - \frac{1}{3}c\right).$$

The same relation is obtained in the first approximation by expansion of the eigenvalue $\mu = (1 - c)/(3 - 2c)$, found for the present problem exactly. This simple example illustrates the basic idea of the renormalization-group-with-perturbation-expansion approach. The following basic points should be noted. A scaling law, in our terms incomplete similarity, is assumed; this scaling law depends on a parameter. For the value zero of the parameter the solution is known. An asymptotic expansion is then used to find the solution for small but finite values of the parameter.

If there is no value of the parameter for which there exists complete similarity then the expansion cannot be performed. The only ways to obtain the 'anomalous dimensions', $\alpha_{\ell+1}, \ldots, \delta_\ell$, are to solve the nonlinear eigenvalue problem or to perform an experiment, physical or numerical.

Chapter 6

Self-similar phenomena and travelling waves

6.1 Travelling waves

In various problems in applied mathematics an important role is played by *travelling waves*. These are phenomena for which distributions of the properties of motion at different times can be obtained from one another by a translation, rather than by a similarity transformation as in the case of self-similar phenomena. In other words, one can always choose a moving Cartesian coordinate system such that the distribution of properties of a phenomenon of travelling-wave type is stationary in that system.

In accordance with the definition given above, solutions of travelling-wave type can be expressed in the form

$$\mathbf{v} = \mathbf{V}(x - X(t)) + \mathbf{V}_0(t). \tag{6.1}$$

Here $\mathbf{v}$ (generally speaking, a vector) is the property of the phenomenon being considered; x is the spatial Cartesian coordinate, an independent variable of the problem; t is another independent variable, for definiteness identified with time, although this is not necessary, and $X(t)$ and $\mathbf{V}_0(t)$ are time-dependent translations along the x- and $\mathbf{v}$-axes. In particular, if the properties of the process do not depend directly on time, so that the equations governing the process do not contain time explicitly, the travelling wave propagates uniformly:

$$\mathbf{v} = \mathbf{V}(x - \lambda t + c) + \mu t. \tag{6.2}$$

Here λ, μ and c are constants; c is the phase shift and λ and μ represent the speeds of translation along the x- and $\mathbf{v}$-axes. For an important class of waves, steady travelling waves, the distribution of properties in a wave remains unchanged in time, so that $\mu = 0$ and

$$\mathbf{v} = \mathbf{V}(x - \lambda t + c). \tag{6.3}$$

In particular, steady travelling waves describe the fine structure of 'fronts', which are associated with shock waves, flames and analogous regions of rapidly changing density, speed and other properties of the motion and are described by surfaces of discontinuity when dissipative processes are neglected.

Travelling waves are closely connected with self-similarities. Indeed, if in (6.1) we set

$$\mathbf{v} = \ln \mathbf{u}, \quad t = \ln \tau, \quad \mathbf{V}_0(t) = \ln \mathbf{u}_0(\tau),$$
$$\mathbf{V} = \ln \mathbf{U}, \; x = \ln \xi, \quad X(t) = \ln \xi_0(\tau) \tag{6.4}$$

then we obtain a representation of a travelling wave in the self-similar form

$$\mathbf{u} = \mathbf{u}_0(\tau)\mathbf{U}(\xi/\xi_0(\tau)). \tag{6.5}$$

In particular, the relation (6.2) for a uniformly propagating travelling wave reduces to a self-similar form with scaling self-similar variables:

$$\mathbf{u} = \mathbf{B}\tau^\mu \mathbf{U}(\xi/(A\tau^\lambda)); \tag{6.6}$$

where A and $\mathbf{B}$ are constants.

The simple connection noted here between self-similar solutions and travelling waves is well known; it has been used to simplify the study of some self-similar solutions (see for example Staniukovich 1960). Surprisingly, however, the connection between the classification of self-similar solutions and the well-known classification of steady travelling waves long remained unnoticed.

One distinguishes two types of the fronts mentioned above. For fronts of the first kind (shock waves, detonation waves etc.) the speed of propagation of the front is found from the conservation laws of mass, momentum and energy only. The structure of such a front is adapted to the conservation laws in the sense that for a particular speed of propagation of the front, dictated by the conservation laws, the structure can depend on the character of the dissipative processes in the transition region and the magnitudes of the dissipative coefficients. Of course, analysis of a structure of shock waves allows one to reject unrealizable situations such as shock waves of rarefaction, for which it is impossible to construct a structure but basically the speed of propagation of the front is determined independently of the structure of the transition process.

For fronts of the second kind (a flame, gaseous discharge etc.) the conservation laws become insufficient for determination of the speed of the front: this is found as an eigenvalue in the course of determining the structure of the front, i.e. of determining a solution of travelling-wave type of the equations describing the dissipative processes in the transition region.

It turns out that this classification of travelling waves corresponds exactly to the classification of self-similar solutions discussed above. Here we will consider the simplest examples of steady travelling waves of both types, after which we shall see how the two classifications correspond.

6.2 Burgers' shock waves – steady travelling waves of the first kind

Burgers' equation

$$\partial_t v + v \partial_x v = \nu \partial_{xx}^2 v \tag{6.7}$$

is a simplified mathematical model of the motion of a viscous compressible gas. Here v is the speed, ν the kinematic viscosity, x the spatial coordinate, and t the time. If the viscous term is neglected then (6.7) assumes the form of the simplest model equation of gas dynamics,

$$\partial_t v + v \partial_x v = 0. \tag{6.8}$$

This equation has a solution of uniformly propagating shock-wave type, $v = V(\zeta)$, $\zeta = x - \lambda t + c$, where $V(\zeta)$ is a step function equal to v_1 for $\zeta > 0$ and equal to v_2 for $\zeta \leq 0$, with $v_1 < v_2$. The value of the speed of propagation $\lambda = \lambda_0$ is obtained from the law of conservation of momentum at the front of the discontinuity, which corresponds to (6.8):

$$-\lambda_0(v_1 - v_2) + \frac{v_1^2 - v_2^2}{2} = 0, \tag{6.9}$$

whence we find

$$\lambda_0 = \frac{v_1 + v_2}{2}. \tag{6.10}$$

We now take into account the dissipative process, that is, the viscosity, and return to (6.7). We construct a solution of Burgers' equation of travelling-wave type, $v = V(\zeta)$, $\zeta = x - \lambda t + c$. Substituting this expression for v into (6.7), we have

$$-\lambda \frac{dV}{d\zeta} + V \frac{dV}{d\zeta} = \nu \frac{d^2V}{d\zeta^2}, \tag{6.11}$$

whence, integrating and using the condition $V = v_1$ at $\zeta = \infty$, we find

$$\nu \frac{dV}{d\zeta} = -\lambda(V - v_1) + \frac{V^2 - v_1^2}{2}. \tag{6.12}$$

To satisfy the condition at the left-hand endpoint, $V(-\infty) = v_2$, it is necessary to take

$$\lambda = \frac{v_1 + v_2}{2} = \lambda_0, \tag{6.13}$$

after which a solution is obtained in the form

$$\frac{\zeta}{\nu} = \frac{2}{v_2 - v_1} \ln \frac{v_2 - V}{V - v_1}. \tag{6.14}$$

This solution describes the structure of the transition region on the length scale $\nu/(v_2 - v_1)$ characteristic of this region. We see that the condition $v_2 > v_1$ imposed above is essential, since a solution describing the structure of the transition region of a wave with $V(-\infty) = v_2 < v_1 = V(\infty)$ does not exist. In fact, with (6.13) taken into account, (6.12) assumes the form

$$\nu \frac{dV}{d\zeta} = -\frac{(v_2 - V)(V - v_1)}{2}. \tag{6.15}$$

Since V lies between v_1 and v_2, the right-hand side of (6.15) is always negative, and the left-hand side is negative only for $v_2 > v_1$.

A solution of travelling-wave type with $\lambda = \lambda_0$ serves as an asymptotic representation of a solution of an initial-value problem for Burgers' equation with initial data of transitional type,

$$\begin{aligned} v(x, 0) &\equiv v_2, & x \leq a, \\ v_1 < v(x, 0) &< v_2, & a < x < b, \\ v(x, 0) &\equiv v_1, & x \geq b, \end{aligned} \tag{6.16}$$

where a and b ($a < b$) are arbitrary real numbers and the function $v(x, 0)$ is monotonically non-increasing, $\partial_x v(x, 0) \leq 0$. This was proved by Oleynik (1957). As is evident, in the present case the value of the speed of propagation λ_0 is obtained from a conservation law and is independent of the structure of the wave, i.e. of the viscosity ν. As (6.14) shows, the viscosity determines only the spatial scale of the transition region, i.e. the 'width' of the front.

The situation is completely analogous for shock waves in gases and detonation waves: the speed of propagation of these waves is determined from the laws of conservation of mass, momentum and energy alone and does not require for its determination any consideration of the wave structure. The latter determines only the width of the transition region.

6.3 Flames: steady travelling waves of the second kind. Nonlinear eigenvalue problem

6.3.1 Schematic formulation of the flame-propagation problem

We now consider travelling waves of the second kind, for which the speed of propagation cannot be found from conservation laws alone but is determined by analysis of the wave structure.

Travelling waves are fundamental intermediate asymptotics in the theory of *flame propagation*. The phenomenon of flame propagation is of enormous fundamental and practical importance. Theoretical studies were begun by Taffanel (1913, 1914) and Daniell (1930) and found an ultimate formulation in the works of Zeldovich and Frank-Kamenetsky (1938a, b); see also Zeldovich (1948).

Here we will present a simplified qualitative model of flame propagation in gaseous mixtures; a more detailed and general discussion of the phenomenon can be found in the book Zeldovich *et al.* (1985). We consider thermal flame propagation in long pipes with thermally isolated walls. Assume that an exothermic chemical reaction is proceeding in a gaseous mixture filling a long pipe and that in the course of this reaction a combustible component of the mixture whose concentration we denote by n is annihilated. The reaction rate Φ, i.e. the mass of combustible matter annihilated in unit volume in unit time, depends on the concentration n and the temperature T:

$$\Phi = \Phi(n, T). \tag{6.17}$$

It is known from physical chemistry that the temperature dependence of reaction rates in flames is very strong: a small change in temperature greatly changes the reaction rate. As a characteristic example the Arrhenius-type dependence can be used:

$$\Phi = An^p \, e^{-E/(RT)}. \tag{6.18}$$

Here A, the reaction order p and the activation energy E are constants and R is the universal gas constant, equal to 2 cal/(mol K). The activation energy for combustion reactions has a typical value of 40 kcal/mol. So, for example, at a room temperature of 300 K the factor $e^{-E/(RT)}$ is equal to approximately 10^{-30} while for a temperature of 1000 K it equals approximately 10^{-9}: thus, a tripling of temperature leads to an increase in the reaction rate by a factor 10^{21}. Therefore Zeldovich (1948) made the key assumption that the function $\Phi(n, T)$ vanishes not only in the original state of the gaseous combustible mixture, when $T = T_1$, but also in a certain temperature interval $T_1 \leq T \leq T_1 + \Delta, \Delta > 0$,

lying above the initial temperature. Thus it is assumed that the function $\Phi(n, T)$
satisfies the conditions

$$\Phi(n, T) \geq 0;$$
$$\Phi(n, T) = 0, \qquad 0 \leq n \leq 1, \qquad T_1 \leq T \leq T_1 + \Delta; \qquad \Phi(0, T) = 0. \quad (6.19)$$

For our qualitative model we will assume that the pressure and density ρ of
the gas mixture remain constant and we will neglect the gas motion. Then the
equations of balance of energy and of the combustible component of the gas
mixture assume the forms

$$\partial_t n = D\partial^2_{xx} n - \frac{1}{\rho} \Phi(n, T), \tag{6.20}$$

$$\partial_t T = \kappa \partial^2_{xx} T + \frac{Q}{\rho c_p} \Phi(n, T). \tag{6.21}$$

Here D is the coefficient of diffusion, $\kappa = k/(\rho c_p)$ is the thermal diffusivity,
k is the thermal conductivity, c_p is the specific heat at constant pressure and
Q is the thermal efficiency of the exothermic reaction, the amount of heat
generated by burning a unit of mass of the combustible substance. All these
quantities are assumed to be constant. From physical chemistry it is known that
if the combustible mixture and the combustion products have close molecular
weights then the coefficient of diffusion D is close to the coefficient of thermal
diffusivity κ. We will assume that $D = \kappa$, so multiplying (6.20) by Q/c_p and
adding to (6.21) we obtain for the quantity $T + Qn/c_p$ the linear diffusion
equation:

$$\partial_t \left(T + \frac{Q}{c_p} n \right) = \kappa \partial^2_{xx} \left(T + \frac{Q}{c_p} n \right). \tag{6.22}$$

The combustion zone is small in comparison with the length of the pipe, which
we can therefore consider as infinite, so that $-\infty < x < \infty$. Assume further
that, at the beginning, half the tube $(x > 0)$ is filled by fresh mixture, i.e.
$n = 1, T = T_1$ for $x > 0$, and the other half $(x < 0)$ contains the products of
combustion, i.e. $n = 0, T = T_1 + Q/c_p = T_2$ for $x < 0$. This means that, at the
beginning, everywhere in the tube the quantity $T + Qn/c_p$ is a constant equal
to the temperature of the combustion products T_2:

$$T + \frac{Q}{c_p} n = T_1 + \frac{Q}{c_p} = T_2. \tag{6.23}$$

In the theory of combustion the relation (6.23) is called the Lewis–von Elbe
similarity law for the concentration and temperature fields. This similarity law
allows the exclusion of the concentration from the expression for the reaction

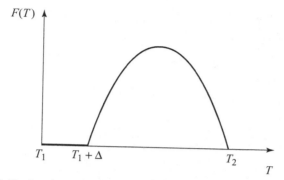

Figure 6.1. The function $F(T)$ vanishes in some interval near $T = T_1$ in the problem of flame propagation.

rate, so that we can write

$$\Phi = \Phi\left(\frac{(T_2 - T)c_p}{Q}, T\right).$$

$$(6.24)$$

The equation (6.21) then takes the form

$$\partial_t T = \kappa \partial_{xx}^2 T + F(T),$$

$$(6.25)$$

where

$$F(T) = \frac{Q}{\rho c_p} \Phi\left(\frac{(T_2 - T)c_p}{Q}, T\right).$$

$$(6.26)$$

The function $F(T)$ is by assumption identically equal to zero near the temperature of the fresh mixture: $T_1 \leq T \leq T_1 + \Delta$ (see Figure 6.1). As a model of a flame, the intermediate asymptotic solution of equation (6.25) is considered to be of travelling-wave type:

$$T = T(\zeta), \qquad \zeta = x - \lambda t + c.$$

$$(6.27)$$

Here λ is the speed of propagation of the travelling wave and c is a constant phase which is not obtained in the process of direct construction of the travelling-wave solution.

Substituting (6.27) into (6.25) we obtain for the function T an ordinary differential equation:

$$\lambda \frac{dT}{d\zeta} + \kappa \frac{d^2 T}{d\zeta^2} + F(T) = 0.$$

$$(6.28)$$

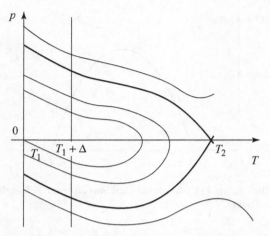

Figure 6.2. Phase portrait of the integral curves of the first-order equation (6.29).

The derivative $dT/d\zeta = p$ is proportional to the heat flux. Taking T as an independent variable and taking for simplicity $x = 1$, we obtain for p a first-order equation:

$$p\frac{dp}{dT} + \lambda p + F(T) = 0. \tag{6.29}$$

The heat flux vanishes both at $\zeta = \infty$ where $T = T_1$ and at $\zeta = -\infty$ where $T = T_2$. Therefore there are two boundary conditions for the first-order equation (6.29):

$$p = 0, \quad T = T_1; \qquad p = 0, \quad T = T_2. \tag{6.30}$$

6.3.2 Nonlinear eigenvalue problem

We have again, as in the case of self-similar solutions of the second kind, a nonlinear eigenvalue problem: (6.29) is a first-order equation and (6.30) gives us two boundary conditions. We will show, following Zeldovich (1948), that there exists a unique eigenvalue λ for which the desired solution exists. We consider the phase portrait of (6.29) in the region of interest to us in the T p-plane (Figure 6.2). At $T = T_2$ and $p = 0$, (6.29) has a singular point of saddle type. Through this singular point pass two separatrices with slopes $-\lambda/2 \pm [\lambda^2/4 - F'(T_2)]^{1/2}$; since $F'(T_2) < 0$, the slope of one of the separatrices is positive and of the other is negative. It is clear that only the separatrices can satisfy the second condition of (6.30). Furthermore, for $\lambda = 0$, (6.29) can be integrated in finite form: the solutions satisfying the condition (6.30) for $T = T_2$

have the form

$$p = \pm \left\{ 2 \int_T^{T_2} F(T)\,dT \right\}^{1/2}, \qquad (6.31)$$

so that the ordinates of the points of intersection of the corresponding integral curves with the vertical axis are

$$p_1 = \left\{ 2 \int_{T_1}^{T_2} F(T)\,dT \right\}^{1/2} > 0, \qquad p_2 = -\left\{ 2 \int_{T_1}^{T_2} F(T)\,dT \right\}^{1/2} < 0. \qquad (6.32)$$

We now consider the function $q(T, \lambda) \equiv \partial_\lambda p$ for all solutions of (6.29) satisfying the second condition (6.30). It is clear that $q(T_2, \lambda) \equiv 0$ since $p(T_2, \lambda) \equiv 0$. Differentiating (6.29), we obtain for the function q the equation

$$\frac{dq}{dT} = -1 + \frac{F(T)}{p^2}\,q. \qquad (6.33)$$

Close to the point $T = T_2$ the separatrices behave, according to the above, like $p = (T - T_2)\{-\lambda/2 \pm [\lambda^2/4 - F'(T_2)]^{1/2}\}$. Differentiating with respect to λ, we find that the corresponding curves $q(T, \lambda)$ behave near to $T = T_2$ like $q = K(T - T_2)$, where

$$K = \left\{ -\frac{1}{2} \pm \frac{\lambda}{2\sqrt{\lambda^2 - 4F'(T_2)}} \right\}$$

is negative for both separatrices, i.e. $q > 0$ for $T < T_2$. Furthermore there cannot be an intersection of the curve $q(T, \lambda)$ with the axis $q = 0$ at some point intermediate between T_1 and T_2, because at a point of intersection one would have $dq/dT = -1$, which is geometrically impossible. Thus, $q(T_1 + \Delta, \lambda) > 0$. But for $T_1 \leq T \leq T_1 + \Delta$ we have $F(T) \equiv 0$, and from this and (6.33) we get $q(T_1, \lambda) = q(T_1 + \Delta, \lambda) + \Delta > \Delta$. Since

$$p(T_1, \lambda) = p(T_1, 0) + \int_0^\lambda q(T_1, \lambda)\,d\lambda > -\left\{ 2 \int_{T_1}^{T_2} F(T)\,dT \right\}^{1/2} + \lambda\Delta, \qquad (6.34)$$

it follows that one can find one and only one value $\lambda = \lambda_0$ such that the lower separatrix reaches the point $p = 0$, $T = T_1$, i.e. satisfies all the conditions of the problem.

Thus, the existence and uniqueness of the solution of the nonlinear eigenvalue problem is proved. The construction of the solution presented above can easily be performed numerically.

In fact, the investigation of travelling-wave solutions of the equations of the type (6.25) began with a rigorous mathematical study, undertaken in the fundamental work of Kolmogorov, Petrovskii and Piskunov (1937). This work was carried out in connection with a biological problem concerning the speed of propagation of a gene that has an advantage in the struggle for life. A remarkable study of this phenomenon was developed independently and simultaneously by Fisher (1937); see also the book Murray (1977). To describe the structure of the transition zone near the boundary of the domains of habitation of genes of both types (advantaged and disadvantaged) they obtained a nonlinear diffusion equation of the same type as (6.25):

$$\partial_t v = \kappa \partial_{xx}^2 v + F(v), \tag{6.35}$$

where v is the gene concentration and $F(v)$ is a continuous function that is differentiable the necessary number of times, defined in the interval $0 \leq v \leq 1$ and having, in accordance with the physical meaning of the problem, the following properties:

$$F(0) = F(1) = 0; \qquad F(v) > 0, \quad 0 < v < 1;$$
$$\tag{6.36}$$
$$F'(0) = \alpha > 0; \qquad F'(v) < \alpha, \quad 0 < v < 1.$$

For the special problem considered by Kolmogorov, Petrovskii, Piskunov and Fisher, $F(v) = \alpha v(1 - v)^2$. Here κ and α are positive constants.

Under these conditions equation (6.35) has a solution of travelling-wave type, $v = V(\zeta)$, $\zeta = x - \lambda t + c$, satisfying the conditions $v(-\infty) = 1$, $v(\infty) = 0$ for all speeds of propagation λ greater than or equal to $\lambda_0 = 2(\kappa \alpha)^{1/2}$ and for arbitrary c. It is of prime importance that among these solutions only that corresponding to the lowest speed of propagation can be an intermediate-asymptotic representation as $t \to \infty$ of solutions of an initial-value problem with conditions of the transitional type:

$$v(x, 0) \equiv 1, \qquad x \leq a,$$
$$0 < v(x, 0) < 1, \qquad a < x < b, \tag{6.37}$$
$$v(x, 0) \equiv 0, \qquad x \geq b.$$

In other words, it turns out that the direct consideration of solutions of travelling-wave type gives a continuous 'spectrum' of possible speeds of propagation $\lambda \geq \lambda_0 = 2(\kappa \alpha)^{1/2}$. However, only the solution corresponding to the lowest point, $\lambda = \lambda_0$, of this spectrum can be an asymptotic solution as $t \to \infty$ of the initial-value problem with conditions of transitional type; the remaining travelling waves are unstable. The quantity λ_0 determines the required speed of propagation of the gene that has an advantage in the struggle for life.

We emphasize that in the gene-propagation problem as well as in the flame-propagation problem, direct construction of a solution of travelling-wave type, $T = T(x - \lambda t + c)$, determines the solution to within a constant phase c. This latter constant can be found only by matching the travelling-wave-type solution with a non-invariant solution of the original problem corresponding to certain initial conditions of transitional type. It is obvious that no matter what intermediate state of the system $T(x, t)$ we take as the initial state, the constant c remains unchanged. In this sense c is an integral (see Lax 1968), but an implicit integral.

We emphasize that there is an essential difference between the problems of gene propagation and flame propagation. In the former, the spectrum – the set of eigenvalues – is continuous, $\lambda \geq \lambda_0 = 2\sqrt{\kappa F'(0)}$, $F'(0) = \alpha > 0$, but only the lowest point of the spectrum corresponds to an intermediate asymptotics. In contrast, in the problem of flame propagation the spectrum consists of a single point. This difference is due to the 'cutting' of the function F.

6.4 Self-similar interpretation of solitons

The self-similar interpretation of solitons is instructive because it demonstrates clearly that the powers in self-similar variables can be arbitrary numbers depending continuously also upon the initial conditions. These numbers can be rational, algebraic, irrational or even transcendental; they are not necessarily the simple fractions which appear when self-similar solutions of the first kind based on integral conservation laws (cf. Chapter 2) are constructed.

Consider the Korteweg–de Vries equation, which appeared initially in the theory of surface waves on shallow water and was later encountered as a qualitative model in numerous other problems:

$$\partial_t u + u \partial_x u + \beta \partial^3_{xxx} u = 0. \tag{6.38}$$

In the theory of surface waves u is, to within a constant factor, the horizontal velocity component and is constant, in the present approximation, over the channel depth; $\beta = c_0 h^2 / 6$, $c_0 = (gh)^{1/2}$, g is the acceleration of gravity, h is the undisturbed depth of the fluid layer, t is the time and x is the horizontal coordinate in a system moving with speed c_0 relative to the fluid at rest at infinity. An analogous equation is valid also in the corresponding approximation for the elevation of the free surface over its undisturbed level. Equation (6.38) has solutions of solitary travelling-wave type, the so-called solitons (Figure 6.3), which are given by

$$u = \frac{u_0}{\cosh^2[\sqrt{u_0/(12\beta)}\,\zeta]} \tag{6.39}$$

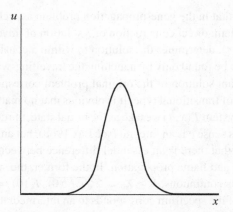

Figure 6.3. A solitary wave, known as a soliton.

where $\zeta = x - \lambda t + c$ and $u_0 = 3\lambda$. (The name 'soliton' reflects the particle-like behaviour of such a solution: after a 'collision' it remains the same except that its 'phase' c changes, generally speaking.)

The solution (6.39) satisfies the conditions

$$u(\infty) = u(-\infty) = 0 \qquad (6.40)$$

for any $\lambda > 0$; the spectrum of eigenvalues λ is continuous and semi-bounded; $\lambda \geq 0$. There is, however, an essential difference between the continuous spectrum in the problem of gene propagation considered in the previous section and the continuous spectrum in this problem. In the former problem, only the lowest point $\lambda = \lambda_0$ of the spectrum satisfies the requirement that the solution of the initial-value problem with initial data of transitional type tends to the given solution of travelling-wave type as $t \to \infty$; for all other λ this is not so, and therefore the corresponding solutions are unstable. For the Korteweg–de Vries equation a remarkable discovery was made by Gardner, Greene, Kruskal and Miura (1967): as $t \to \infty$ for large positive x the solution of the initial-value problem, for initial data $u(x, 0)$ that decrease sufficiently rapidly at $x = \pm\infty$, is represented asymptotically (Figure 6.4) by a finite sum of solutions of the form (6.39):

$$u \sim \sum_{n=1}^{N} 2|\mu_n| \cosh^{-2} \left\{ \sqrt{\frac{|\mu_n|}{6\beta}} \left(x - \frac{2|\mu_n|t}{3} + c_n \right) \right\}, \qquad (6.41)$$

where the μ_n are the discrete eigenvalues of the Schrödinger operator, well known from quantum mechanics, with the potential set equal to $-u(x, 0)$:

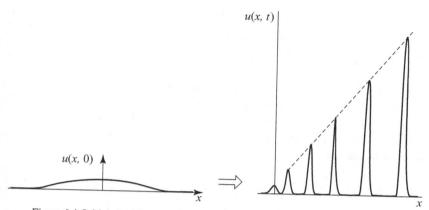

Figure 6.4. Initial elevation of the free surface of a heavy fluid in a shallow channel generates a finite series of solitary waves (solitons).

$$\frac{d^2\Psi}{dx^2} + \frac{1}{6\beta}[\mu + u(x,0)]\Psi = 0, \qquad \Psi(\pm\infty) = 0, \qquad (6.42)$$

and the 'phases' c_n are certain constants, also determined by the initial condition. Hence any solution of soliton type can be an intermediate asymptotics of the solution of an initial-value problem as $t \to \infty$, but exactly which one it is will be determined by the initial conditions. The general meaning of this result was elucidated in a fundamental paper by Lax (1968).

There is an instructive self-similar interpretation of the result (6.41) presented above for the Korteweg–de Vries equation (6.38). If we set $x = \ln \xi$, $t = \ln \tau$ then equation (6.38) can be rewritten in the form

$$\tau \partial_\tau u + \xi u \partial_\xi u + \beta \left(\xi^3 \partial^3_{\xi\xi\xi} u + 3\xi^2 \partial^2_{\xi\xi} u + \xi \partial_\xi u\right) = 0. \qquad (6.43)$$

The solution of travelling-wave type (6.39) here assumes the self-similar form

$$u = \frac{12\lambda}{2 + \eta^{\sqrt{\lambda/\beta}} + \eta^{-\sqrt{\lambda/\beta}}}, \qquad \eta = \frac{\xi}{A\tau^\lambda}. \qquad (6.44)$$

Here $A = e^{-c}$ is constant. We note that the right-hand side of (6.44) is not small only for η of order unity; it is small if η is either large or small. The spectrum of eigenvalues λ, obtained by direct construction of solutions of travelling-wave type, is continuous and semi-bounded: $\lambda \geq 0$. The result of Gardner, Greene, Kruskal and Miura (1967) presented above can be expressed in a self-similar interpretation in the following way: an asymptotic solution of the initial-value problem for (6.43) as $\tau \to \infty$ and for large ξ can be represented in the form

$$
u \sim \sum_{n=1}^{N} 12\lambda_n \left\{ 2 + \left(\frac{\xi}{A_n \tau^{\lambda_n}} \right)^{\sqrt{\lambda_n/\beta}} + \left(\frac{\xi}{A_n \tau^{\lambda_n}} \right)^{-\sqrt{\lambda_n/\beta}} \right\}^{-1}. \tag{6.45}
$$

Thus the initial distribution $u(\xi, 0)$, which by assumption decreases sufficiently rapidly as $\xi \to 0$ or ∞, determines another N positive constants $\lambda_1, \ldots, \lambda_N$ and another N positive constants $A_1, \ldots, A_N$ and selects N intervals in ξ. Inside each interval $\xi = O(\tau^{\lambda_n})$, the asymptotics of the solution is self-similar and has the form

$$
u \sim 12\lambda_n \left\{ 2 + \left(\frac{\xi}{A_n \tau^{\lambda_n}} \right)^{\sqrt{\lambda_n/\beta}} + \left(\frac{\xi}{A_n \tau^{\lambda_n}} \right)^{-\sqrt{\lambda_n/\beta}} \right\}^{-1}. \tag{6.46}
$$

Outside the intervals mentioned the solution u is small: $u = o(1)$. Here it is significant that in the self-similar asymptotics not only do the constants A_n depend as usual upon the initial conditions of the original non-idealized problem but *so also do the powers λ_n in the expressions for the self-similar variables.* Variations in the initial condition lead to a continuous variation in the powers. The constants A_n and λ_n are the implicit integrals of the initial-value problem.

Chapter 7
Scaling laws and fractals

7.1 Mandelbrot fractals and incomplete similarity

7.1.1 The concept of fractals. Fractal curves

In the scientific and even popular literature of recent time *fractals* have been widely used and discussed. By fractals are meant those geometric objects, curves, surfaces and three- and higher-dimensional bodies, having a rugged form and possessing certain special properties of homogeneity and self-similarity. Such geometric objects were studied intensively by mathematicians at the end of the nineteenth century and the beginning of the twentieth century, euphony particularly in connection with the construction of examples of continuous nowhere-differentiable functions. To many pure mathematicians (starting with Hermite) and most physicists and engineers they seemed for a long time mathematical monsters having no applications in the problems of natural science and technology. In fact, it is not so and in clarifying this point the concept of intermediate asymptotics plays a decisive role.

The revival of interest in such objects and the recognition of their fundamental role in natural science and engineering is due primarily to a series of papers by Mandelbrot and, especially, to his monographs (1975, 1977, 1982). Mandelbrot coined the very term 'fractal' and introduced the general concept of fractality. In the monographs and subsequent papers Mandelbrot and his followers showed that, contrary to what was expected, this concept, enclosing many special examples known before, appeared to be fruitful in such diverse and important applications as polymer physics, geomorphology, the theory of Brownian motion, turbulence theory, astrophysics, fracture theory and many others. In the monographs of Mandelbrot are presented from a unified viewpoint the previous works of other authors that relate to these topics.[1]

In this chapter we will demonstrate the concept of Mandelbrot fractals, using first the simplest example, fractal curves. We will discuss the properties of homogeneity and self-similarity that make a continuous curve fractal, and we will show that the very idea of fractals is closely related to the incomplete-similarity concept. A non-trivial example related to the fractality of respiratory organs will be presented in conclusion.

We will start from an instructive example. The English physicist L.F. Richardson (see Richardson 1961; Mandelbrot 1975, 1977, 1982) was commissioned to determine the length of the west coast of Britain. Richardson chose the following way of solving this problem, which would be quite natural for ordinary smooth curves. He approximated the coastline on the most detailed available map of Britain by a broken line composed of segments of constant length η, all vertices of which were situated on the coastline. The length L_η of this broken line was taken as the approximate value of the coastline's length corresponding to a particular value of η. Richardson assumed at first that, when reducing η, the values L_η of the length of the approximating broken line will tend to a definite finite limit, which should be considered as the coastline's length.

Naturally, this is found to be the case when this method is used for a circle (Figure 7.1(a)). However, the west coastline of Britain turned out to be so rugged, even down to the smallest scales available on the map, that the value L_η did not tend to a finite limit as the segment length η of the approximating broken line was reduced. Just the opposite: the value L_η tended to infinity as η tended to zero; throughout the available range of η the growth in L_η followed the scaling law (Figure 7.1(b))

$$L_\eta = \lambda \eta^{1-D}, \tag{7.1}$$

where $\lambda > 0$ and $D, 2 > D > 1$, are certain constants. For the approximate lengths of separate parts of the same coastline between certain points of it relations of the form (7.1) were again obtained, with the same D but a different, smaller, value of λ. When such a procedure was performed for the coastline of Australia (Figure 7.1(c)) the scaling law remained, but this time both λ and D were found to be different. As can be seen, D is dimensionless; however, λ has the unusual dimension of length raised to a non-integer fractional power.

---

[1] Mandelbrot's success was so complete that nowadays people try to find fractals everywhere. I have to emphasize therefore that fractals in their turn are very special non-generic objects.

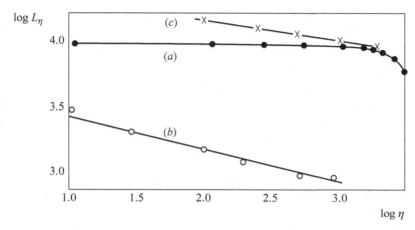

Figure 7.1. The dependence of the length L_η of a broken line on the segment length η for (*a*) a circle, (*b*) the west coast of Britain, (*c*) the Australian coast. (After Mandelbrot 1977.)

Formal passage to the limit $\eta \to 0$ in relation (7.1) gives a result rather unexpected in its content: the length of the coastline determined by the method proposed, and even the length of each part of it, appears to be infinite. The substantial point here is that if one tries to use a more detailed map in the hope that there the desired limit will appear, he or she will discover that such a map is somewhat meaningless because due to tides the very concept of the coastline is restricted to rather large scales.

It follows from (7.1) that parts of the coastline can be compared by a certain measure of their extent, although not by their length. In fact, let us approximate two pieces of coastline by broken lines with the same segment length η. In both cases relations of the form (7.1) are obtained:

$$L_\eta^{(1)} = \lambda^{(1)}\eta^{1-D}, \qquad L_\eta^{(2)} = \lambda^{(2)}\eta^{1-D}. \tag{7.2}$$

As is seen, the ratio $L_\eta^{(2)}/L_\eta^{(1)} = \lambda^{(2)}/\lambda^{(1)}$ does not depend on the segment length η. Therefore, the extent of certain parts of the coastline can be compared, not however by their lengths but by the corresponding coefficients λ. Thus, the very approach of measuring the extent of the coastline by the same means as for smooth curves is found to be inapplicable.

Richardson found an adequate image of the west coast of Britain in a curve of a different type. To understand this type of curve we consider first circle (Figure 7.2) and inscribe in the circle a regular n-gon with side length η. The perimeter of this inscribed polygon L_η obviously depends only on the diameter

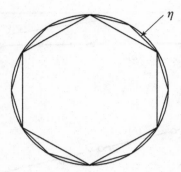

Figure 7.2. A circle with inscribed regular polygons. As the number of sides in the polygon approaches infinity, and the side length η approaches zero, the perimeter of the polygon approaches a finite limit.

of the circle d and the side length of the polygon η:

$$L_\eta = f(d, \eta). \tag{7.3}$$

Both arguments of f have the dimension of length. Using dimensional analysis in the traditional way we can transform (7.3) to the form

$$\Pi = \Phi(\Pi_1), \tag{7.4}$$

where $\Pi = L_\eta/d$ and $\Pi_1 = \eta/d$, whence

$$L_\eta = d\Phi\left(\frac{\eta}{d}\right). \tag{7.5}$$

Let the number of sides of the polygon n approach infinity, i.e. let the side length η approach zero. From elementary geometry, it is known that the perimeter of the inscribed polygon approaches the finite limit $L_0 = \pi d$ (which is, in fact, adopted as the circumference of a circle). Thus, as $\eta/d \to 0$ the function $\Phi(\eta/d)$ approaches a finite limit equal to π. Therefore, for sufficiently small η/d it is possible to neglect the influence of the parameter η and to assume that the following relation is satisfied to the required accuracy for polygons with a very large number of sides:

$$\Pi = \text{const} = \pi, \tag{7.6}$$

i.e. $L_\eta = \pi d$.

The second curve is obtained in the following way (Figure 7.3). An equilateral triangle of side d is taken, and each of its three sides is subjected to the following *elementary operation*: the side is divided into three sections, and the middle section is replaced by two sides of an equilateral triangle, constructed using it as a base. The sides of the polygon obtained are once again subjected to the same elementary operation, and so on to infinity. Obviously, the side length of

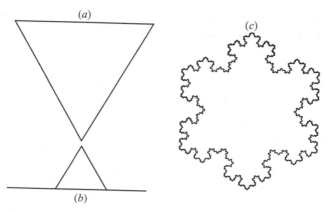

(a)

(c)

(b)

Figure 7.3. A fractal curve – the Koch triad. (a) The original triangle, (b) the elementary operation and (c) the broken line that approximates the fractal curve for a large number of sides. As the number of sides increases, the perimeter of the broken line approaches infinity according to a power law.

this polygon at the nth stage, η, is equal to $d/3^n$, and the perimeter of the entire polygon, L_η, is equal to $3d(4/3)^n$. Equations (7.4) and (7.5) clearly also hold in this case. However, it can easily be shown that, since

$$n = \frac{\log(d/\eta)}{\log 3},$$

we have

$$L_\eta = 3d\left[10^{n(\log 4 - \log 3)}\right] = 3d\left[10^{\alpha \log(d/\eta)}\right] = 3d(d/\eta)^\alpha \qquad (7.7)$$

where

$$\alpha = (\log 4 - \log 3)/\log 3 \simeq 0.26\ldots.$$

Comparing (7.7) and (7.5), we find that

$$\Phi(\eta/d) = 3(\eta/d)^{-\alpha}, \qquad (7.8)$$

i.e. the length of the curve L_0 is infinite in this case, so that only the empty relation $\Pi = \infty$ is obtained in going to the limit $\eta/d \to 0$. Thus, if one is interested in the perimeter of the polygon for large n, it is not possible to pass to the limit and use a limiting relationship such as (7.6). At the same time, equation (7.4) can be rewritten in the form $\Pi^* = \text{const}$, setting

$$\Pi^* = \frac{L_\eta}{d^{1+\alpha}\eta^{-\alpha}}, \qquad \text{const} = 3. \qquad (7.9)$$

The parameter Π^* is (like Π) a power-law combination of the parameters that determine it. However, the structure of (7.9) is not determined by dimensional considerations alone; we do not know the number α beforehand as we did in the case of a circle. Furthermore, unlike the case of a circle, equation (7.6), the parameter η remains in the resulting equation no matter how small η/d is. Therefore, the length of the inscribed broken line, $L_\eta = 3d^{1+\alpha}/\eta^\alpha$, turns out to be proportional to $d^{1+\alpha}$ rather than d, and the length of a segment of this broken line, η, remains in the constant of proportionality.

The construction of the curve presented in Figure 7.3(c) was performed by von Koch (1904), and this curve is called the von Koch triad.

So, Richardson understood that an adequate representation of the coastline is a curve of the von Koch triad type. Indeed, for such curves the relation obtained by Richardson as an empirical equation, (7.1), is also valid, if we write in (7.7) $\lambda = 3d^{1+\alpha}$, $D = 1 + \alpha$.

It follows from relation (7.1) that the number of segments of length η of the approximating broken line is

$$N_\eta = L_\eta/\eta = \lambda\eta^{-D}. \qquad (7.10)$$

The quantity L_η, the length of the approximating broken line, according to (7.7) tends to infinity as $\eta \to 0$, because $D > 1$. Let us construct a square on each segment of the approximating broken line. The total area of these squares is equal to $N_\eta\eta^2 = \lambda\eta^{2-D}$. This quantity tends to zero as $\eta \to 0$, because $D < 2$. Therefore, roughly speaking, the length of this curve is infinite and its area is equal to zero. However, a finite quantity, different from zero, is obtained in the limit $\eta \to 0$ if the number of segments in the approximating broken line is multiplied by η raised to a power D, intermediate between 1 and 2:

$$N_\eta \eta^D = \lambda. \qquad (7.11)$$

The constant D is called the *fractal dimension* of the curve considered. For the fractal dimension of the von Koch triad the double inequality $1 < D < 2$ is valid. The same follows for the coastlines: for the west coast of Britain $D \simeq 1.24$, and for the Australian coastline $D \simeq 1.13$ (Figure 7.1). Thus, for these curves also the fractal dimension lies between 1 and 2. However, the length of the approximating broken line for ordinary smooth curves is bounded, so for smooth curves $D = 1$. It is clear that the fractal dimension is defined not for all continuous curves but only for those where relation (7.1) for the length of the approximating broken lines holds. Let us now give a formal definition of fractal curves:

A fractal curve is a continuous curve for which the fractal dimension is strictly larger than unity:

$$D > 1. \qquad (7.12)$$

From what has been said it follows that the von Koch triad is a fractal curve. As is shown by Richardson's analysis, presented above, the coastlines of the British west coast and of Australia are also adequately approximated by fractal curves.

Note that constancy of the fractal dimension along the whole fractal curve is not necessary. To be fractal, a continuous curve should allow, in the vicinity of each point, a local approximation of the curve by broken lines whose length is represented by a relation of the type (7.1), where D is in general more than unity but can be different for different points.

The consideration of fractals presented above for the example of fractal curves can be in principle extended very simply to surfaces, volumes and to objects of arbitrary topological dimension. For instance, fractal surfaces should be approximated by surfaces composed of tetrahedrons (see section 7.3).

7.2 Incomplete similarity of fractals

Let us explain the properties of fractal curves considered above which lead to a scaling law for the length of the approximating broken lines when the segment length is reduced. Consider a continuous closed curve, whose diameter (the distance between the furthest points) is equal to d. Approximate the curve considered by a broken line with constant segment length η,[2] its vertices being situated on the curve. It is clear that the number of segments N_η of the broken line depends on the dimensional parameters d and η. The quantity N_η is dimensionless; therefore dimensional analysis gives in a standard way

$$N_\eta = f(d/\eta). \tag{7.13}$$

Let us take another approximating broken line with a smaller length of segment $\xi < \eta$. Consider the portion of the basic curve between two neighbouring vertices of the first broken line and let us attempt to determine the number of vertices of the second curve contained in this portion. The von Koch triad has two very important properties. The first is *homogeneity*: all portions of the basic curve between neighbouring vertices of the first broken line generate equal numbers of segments of the second broken line. The second is *self-similarity* (the similarity of the whole curve to any part of it): the number of segments of the broken line with segment length ξ that are placed between neighbouring vertices of the broken line with segment length η depends only on the ratio η/ξ,

---

[2] Obviously, the last segment will in general have length less than η, but this does not matter for $\eta \to 0$.

not on η and ξ separately. We shall assume that the curve under consideration also possesses the properties of homogeneity and self-similarity.

Now consider the broken line with segment length equal to the diameter of the curve. The number of segments of such a broken line is equal, according to (7.13), to $f(1)$. Thus, each segment of the broken line, equal to the diameter of the curve, contains $f(d/\eta)/f(1)$ segments of the broken line with segment length η. According to the self-similarity property, the analogous expression with d replaced by η, and η by ξ, holds for the number $N_{\xi\eta}$ of segments of a second broken line, with segment length ξ, that are contained between two neighbouring vertices of the broken line with segment length η:

$$N_{\xi\eta} = f(\eta/\xi)/f(1). \tag{7.14}$$

However, owing to the homogeneity of the basic curve the same relation holds for all segments of the broken line with segment length η, whose number is equal to $f(d/\eta)$. Therefore, on the one hand the total number of segments of the second broken line contained in the basic curve will be equal to

$$\frac{f(d/\eta)f(\eta/\xi)}{f(1)}. \tag{7.15}$$

On the other hand, owing to the same formula (7.13) the number of segments of the second broken line contained in the basic curve must also be equal to $f(d/\xi)$. Equating these two relations we obtain a functional equation for the function f:

$$f(x)f(y/x) = f(y)f(1), \tag{7.16}$$

where $x = d/\eta$ and $y = d/\xi$, so that $\eta/\xi = y/x$. We have met the relevant functional equation already, in Chapter 1, equation (1.6). Equation (7.16) is solved in an analogous way, and we obtain its solution in the form

$$f(x) = Cx^D, \tag{7.17}$$

where $C = f(1)$ and D are constants. Bearing in mind that $L_\eta = N_\eta \eta$, we obtain from (7.13) and (7.17)

$$L_\eta = \lambda\eta^{1-D}, \tag{7.18}$$

where $\lambda = Cd^D$, i.e. the relation (7.1). Thus, we have shown that for a continuous closed curve possessing the properties of homogeneity and self-similarity the scaling law (7.1) is valid, D having a constant value over the whole curve. If $D > 1$, the curve is fractal.

However, the requirements of homogeneity and self-similarity are very restrictive, so the set of curves exactly satisfying them is rather narrow. It is unlikely, for instance, that the curves representing the coastline would satisfy this property exactly. We will show that the properties of homogeneity and self-similarity are not necessary for a curve to be fractal: the much weaker properties of *local homogeneity* and *local self-similarity* are sufficient.

The latter properties imply that for every point on a curve possessing them a small vicinity Δ can be found where the curve has the following property. The leading term in the asymptotic representation (as $\eta/\xi \to \infty$) of the number of vertices $N_{\xi\eta}$ of the approximating broken line with segment length ξ between two neighbouring vertices of the broken line with segment length η, depends, as $\eta/\xi \to \infty$, only on the ratio η/ξ. We may assume therefore that, neglecting small quantities, the number of vertices $N_{\xi\eta}$ of the broken line with segment length ξ inside a segment of the broken line with segment length η does not depend on the position of this latter segment within the vicinity Δ or on the values of η and ξ, given that the ratio $\eta/\xi \gg 1$ is held fixed:

$$N_{\xi\eta} = f(\eta/\xi). \tag{7.19}$$

Consider now a third broken line with still smaller segment length $\zeta \ll \xi$. On the one hand, owing to local homogeneity and self-similarity the number of its segments within one segment of length η positioned in the vicinity Δ is equal, again neglecting small quantities, to $f(\eta/\zeta)$. On the other hand it is equal to the product of the number $f(\eta/\xi)$ of segments of length ξ inside a segment of length η times the number of segments of length ζ inside a segment of length ξ. Equating the two expressions we obtain a functional equation for the function f,

$$f(x)f(y/x) = f(y), \tag{7.20}$$

which coincides with equation (1.6). Here $x = \eta/\xi$ and $y = \eta/\zeta$. The solution to this equation is represented by $f(x) = x^D$; cf. (7.17). The value of D can be different for various parts of the basic curve. Going from the number of segments to the length of the broken line we obtain that for the lengths of the approximating broken lines in the vicinity of each point of a continuous curve possessing the properties of local homogeneity and local self-similarity a scaling asymptotic relation is valid,

$$L_\xi = \eta^D \xi^{1-D} + \cdots \tag{7.21}$$

where the ellipses refer to quantities small in comparison with the first term. If $D > 1$ this means that the curve considered is fractal.

We emphasize again that the set of curves having the properties of local homogeneity and local self-similarity is richer than the set of curves of von Koch triad type, which possess the very special properties of complete homogeneity and self-similarity.

Fractals reveal the property of incomplete similarity. Let us show this for the same example, fractal curves. The length of a broken line of segment length ξ that approximates the continuous curve between two of its points a distance η apart depends on the dimensional parameters η and ξ. Dimensional analysis gives

$$L_\eta = \eta \Phi(\eta/\xi). \qquad (7.22)$$

For a smooth (or piecewise smooth) curve, as $\xi \rightarrow 0$, i.e. as $\eta/\xi \rightarrow \infty$, the function Φ tends to a finite limit, $\Phi(\infty)$. By definition the value

$$L_0 = \Phi(\infty)\eta \qquad (7.23)$$

is the length of a portion of a smooth curve between two of its points a distance η apart. For instance, if the curve considered is a semicircle having the segment η as its diameter, $\Phi(\infty) = \pi/2$. Therefore for smooth curves we have complete similarity in the parameter η/ξ as $\eta/\xi \rightarrow \infty$.

For fractal curves a finite limit of the function $\Phi(\eta/\xi)$ as $\eta/\xi \rightarrow \infty$ does not exist; the limit is equal to infinity. However, it follows from the relations (7.21) and (7.22) that as $\eta/\xi \rightarrow \infty$ the function $\Phi(\eta/\xi)$ has a scaling asymptotic representation,

$$\Phi(\eta/\xi) \simeq (\eta/\xi)^{D-1}, \qquad (7.24)$$

i.e. incomplete similarity occurs in the parameter η/ξ as $\eta/\xi \rightarrow \infty$. It is clear also that the fractal dimension D depends on the geometric properties of the curve and cannot be obtained from dimensional considerations.

We note in conclusion that, passing from geometric objects to the physical objects represented by them, we can simply identify fractality with incomplete similarity.

7.3 Scaling relationship between the breathing rate of animals and their mass. Fractality of respiratory organs

Every animal possesses a respiratory organ that absorbs oxygen from the environment. At first sight, the part of the organ that directly assimilates the oxygen may be schematically represented as a line (this will be the case if the respiratory organ consists of one or more whiskers), or a surface or some volume that, like a kidney, contains a multitude of small absorbent sacs separated by pores along which water, or air, containing oxygen moves. (As we shall see later, the actual situation is more complicated.) Thus, the respiratory organ of an animal can be characterized by some specific absorptive capacity β_n, i.e. the mass of oxygen absorbed per unit time per unit length ($n = 1$), per unit area ($n = 2$) or or per unit volume ($n = 3$) of the respiratory organ, respectively. Of course, the specific absorptive capacity β_n may depend on external conditions: the temperature, the composition of the environment, the time of day, the speed at which the animal is moving etc.

Our basic assumption is that the breathing rate of the animal, i.e. the mass R of oxygen that it absorbs per unit time, is determined by the following quantities: the body mass W of the animal, the density ρ of its body and the specific absorptive capacity β_n of its respiratory organ. Hence

$$R = f(W, \rho, \beta_n). \tag{7.25}$$

We now note an important feature: the mass of oxygen absorbed and the body mass of the animal may be measured in independent units. (The reason is that the change in the body mass of the animal due to breathing in and out is small and may be neglected). Thus, we choose the $LMTM_{O_2}$ class of systems of units, in which M_{O_2} is the dimension of the mass of oxygen absorbed; this is, according to what we have just said, effectively independent of the dimension[3] of the body mass of the animal, M.

The dimensions of the parameter R and of the governing parameters W, ρ and β_n are given, as may easily be seen, by the following relations:

$$[R] = M_{O_2}T^{-1}, \qquad [W] = M, \qquad [\rho] = ML^{-3}, \qquad [\beta_n] = M_{O_2}L^{-n}T^{-1}. \tag{7.26}$$

Thus, the number of governing parameters is equal to 3; they all have independent dimensions, and, according to dimensional analysis, the relation (7.25) can be written in the following dimensionless form:

$$\Pi = \frac{R}{\beta_n(W/\rho)^{n/3}} = \text{const.} \tag{7.27}$$

Hence, we have

$$R = AW^\alpha, \qquad A = \text{const } \beta_n\rho^{-\alpha}, \qquad \alpha = n/3, \tag{7.28}$$

i.e. a scaling relationship between the breathing rate of an animal and its body mass. According to the foregoing, if the respiratory organ consists of whiskers then α is equal to $1/3$; if the respiratory organ is a surface then α is equal to $2/3$; finally, if the oxygen absorption occurs in a volume then α is equal to unity.

Biological data (see Figure 7.4 for some instructive examples) indicate that a scaling relationship of the form (7.28) is in good agreement with experiment.

---

[3] In fact, this is another example of using invariance with respect to an additional group of transformations (cf. the Introduction, p. 5, and Chapter 5, p. 106).

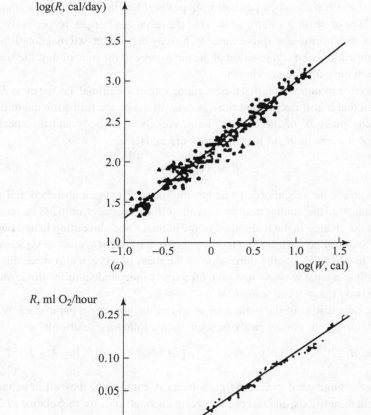

(a)

(b)

Figure 7.4. The rate R of oxygen absorption as a function of body weight W for various animals. (a) Mysids: ▲, sea mysids; ●, farm mysids; ×, laboratory mysids (Shushkina, Kus'micheva and Ostapenko 1971). The rate of oxygen absorption and the body weight of the animals are expressed in energy units, (which are convenient for biologists. The straight line is given by $R = 0.14W^{0.8}$. (b) *Rhithropanopeus harrisii tridentatus* crabs (Nikolaeva 1975). The scaling law is $R = 0.15W^{0.75}$.

From a wide range of data, it is fairly well established that, as a rule, the exponent α lies between 2/3 and unity and very rarely takes on these extreme values.

We interpret this result in the following way: respiratory organs do not have smooth surfaces like a sphere or an ellipsoid but fractal surfaces, i.e. surfaces whose planar cross sections are fractal curves similar to the Koch curve discussed in the previous sections of this chapter.

More precisely, we give the name 'fractal' to a surface that, although continuous, has an extremely broken shape and possesses the property described as follows. We inscribe a polyhedron consisting of triangles with side length η within the surface, just as we inscribed broken lines within the Koch curve. Then, as η tends to zero the total surface area of the polyhedron, S_η, does not approach a finite limit as it does for a smooth surface such as a sphere. On the contrary, S_η goes to infinity according to the scaling law

$$S_\eta = \sigma\eta^{2-D}, \tag{7.29}$$

where σ is some constant having dimension L^D, D being a dimensionless constant greater than 2, but less than 3. The constant D is the *fractal dimension* of the given surface.

Clearly, the area of each face of the inscribed polyhedron is $(\sqrt{3}/4)\eta^2$. From this and (7.29), it follows that the number of faces in the inscribed polyhedron depends on η in the following way: $N_\eta = \mathrm{const}\,\eta^{-D}$.

Thus, for fractal surfaces, the surface area of the inscribed polyhedron tends to infinity as the side length η tends to zero. At the same time, if a prism with altitude η is constructed on each face of the polyhedron then the total volume contained within all such prisms will be $V_\eta = N_\eta(\sqrt{3}/4)\eta^3 = \mathrm{const}\,\eta^{3-D}$; it tends to zero as $\eta \to 0$, since $D < 3$. However, there is some measure of the fractal surface that is intermediate between the area and volume; since the quantity $N_\eta\eta^D$, $2 < D < 3$, approaches a finite limit as η goes to zero, this limit can be used as a measure of the surface mentioned above. Clearly, if the surface of the respiratory organ is a fractal then the specific absorptive capacity of this organ, β_n, should not be defined as the rate of absorption per unit area or volume but per unit of this intermediate dimension. Thus, β_n has dimension

$$[\beta_n] = [R]L^{-D}, \tag{7.30}$$

where D is the fractal dimension of the respiratory organ; D is not restricted to integer values. A comparison of this result with the data presented above (Figure 7.4) and data obtained by other biologists indicates that self-consistency is obtained if one assumes that the respiratory organ is a fractal surface with, for example, $D = 2.4$ for man and sturgeon, $D = 2.4$ for mysids (small sea animals) (Shushkina, Kus'micheva and Ostapenko 1971), $D = 2.25$ for the *Rhithropanopeus harrisii tridentatus* crab (Nikolaeva 1975) etc. The idea that respiratory organs are fractals is also in qualitative agreement with the anatomical data. The analysis performed above was presented by Barenblatt and Monin (1983).

Chapter 8

Scaling laws for turbulent wall-bounded shear flows at very large Reynolds numbers

8.1 Turbulence at very large Reynolds numbers

Turbulence is the state of vortex fluid motion where the velocity, pressure and other properties of the flow field vary in time and space sharply and irregularly and, it can be assumed, randomly. Turbulent fluid flows surround us, in the atmosphere, in the oceans, in engineering systems and biological objects. First recognized and examined by Leonardo da Vinci, for the past century turbulence has been studied by engineers, mathematicians and physicists, including such giants as Kolmogorov, Heisenberg, Taylor, Prandtl and von Kármán. Every advance in a wide collection of subjects, from chaos and fractals to field theory, and every increase in the speed and parallelization of computers is heralded as ushering in the solution of the 'turbulence problem', yet turbulence remains the greatest challenge of applied mathematics as well as of classical physics.

It is very discouraging that in spite of hard work by an army of scientists and research engineers over more than a century, almost nothing became known about turbulence from first principles, i.e. from the continuity equation and the Navier–Stokes equations (Batchelor 1967; Germain 1986; Landau and Lifshitz 1987). These equations are written respectively as

$$\partial_\alpha u_\alpha = 0, \tag{8.1}$$

$$\partial_t u_i + \partial_\alpha u_i u_\alpha = -\frac{1}{\rho} \partial_i p + \nu \Delta u_i. \tag{8.2}$$

Here the standard notation is used: the u_i, $i = 1, 2, 3$, are the velocity components in a rectilinear orthonormal Cartesian coordinate system x_1, x_2, x_3, p is the pressure, t is the time, $\partial_i \equiv \partial/\partial x_i$, Δ is the Laplacian, ν is the kinematic viscosity and ρ is the density; repeated Greek indices imply summation from 1 to 3.

Turbulence at very large Reynolds numbers, often called 'developed turbulence', is widely considered to be one of the happier provinces of the turbulence realm, as it is thought that two of its basic results are well established and should enter, basically untouched, into a future complete theory of turbulence. These results are the von Kármán–Prandtl universal logarithmic law in the wall region of wall-bounded turbulent shear flow and the Kolmogorov–Obukhov scaling laws for the local structure of developed turbulent flow.

The start of fundamental research into turbulent flows at very large Reynolds numbers can be dated sharply from the lecture of Th. von Kármán at the Third International Congress for Applied Mechanics at Stockholm, 25 August 1930. Von Kármán was one of the principal founders of the International Congresses for Applied Mechanics. Unquestionably his lecture 'Mechanical similitude and turbulence' was the central event of the Congress. Von Kármán began his lecture with the following statement:

> Our experimental knowledge of the internal structure of turbulent flows is insufficient for delivering a reliable foundation for a rational theoretical calculation of the velocity distribution and drag in the so-called hydraulic flow state. Numerous semi-empirical formulae, for instance, the attempt to introduce turbulent drag coefficients, are unable to satisfy either the theoretician or the practitioner. The investigations which will be presented below also do not claim to achieve a genuine ultimate theory of turbulence. I will restrict myself rather to clarifying what can be achieved on the basis of pure fluid dynamics if definite hypotheses are introduced concerning definite basic questions.

The hypothesis proposed by von Kármán for answering the fundamental questions concerning the velocity distributions and drag coefficients in turbulent hydraulic flows or, as they are called now, shear flows – first of all, flows in pipes and channels – was presented by him in the following straightforward form:

> On the basis of these experimentally well-established facts we make the assumption that away from the close vicinity of the wall the velocity distribution of the mean flow is viscosity independent.

As a result of subsequent arguments proposed by von Kármán there appeared what is called now the universal (Reynolds-number-independent) logarithmic law and the corresponding drag law for the turbulent flow in a cylindrical pipe. These will be presented below.

The leaders in applied mechanics of that time were present at von Kármán's lecture and took part in the subsequent discussion. The first speaker was L. Prandtl. He said:

> The new Kármán calculations signify very pleasing progress in the problem of fluid friction. It was always the case that by advancing to higher Reynolds numbers the

previous interpolation formulae were revealed to be incorrect by extrapolation to a newly investigated range and had to be replaced by new ones. Research laboratories made big efforts to achieve higher Reynolds numbers, but the cost of big experimental set-ups has a bound which cannot be substantially exceeded. *Due to Kármán's formulae further efforts in this direction became unnecessary* [present author's italics]. The formulae are in such good agreement with the experiments in pipe flows by Nikuradze, and by Schiller and Hermann, and with experiments concerning the drag of plates performed by Kempf, that complete confidence can be placed in them for their application at arbitrarily large Reynolds numbers. For lower Reynolds numbers the agreement is worse, and this can be attributed to the action of the viscosity also in the inner part of the flow, i.e. to the viscosity-influenced streaks of which the laminar layer at the wall consists and which in this case enter far into the internal part of the flow.

I want to point out a seeming contradiction concerning the representation of the velocity distribution by Nikuradze in connection with Kármán's new formulae and my earlier formulation using the dimensionless distance from the wall. Kármán's formulae use viscosity in the boundary condition only. The velocity distribution should be calculated without viscosity. However, the dimensionless distance from the wall, $y^* = (y/\nu)\sqrt{\tau_0/\rho}$, does contain the viscosity. According to my opinion, the explanation is that the Kármán representation should be considered as exact for very large Reynolds numbers, while the representation via the dimensionless distance from the wall applies essentially to the wall layer and streaks where the viscosity and turbulence are acting together.

It should be understood that at that time Ludwig Prandtl was generally considered as 'the chief of applied mechanicians' (cf. Batchelor 1996, p. 185). The opinion which we have just reproduced explains at least partially why over nearly 70 years the Nikuradze (1932) experiments were never extended to larger Reynolds numbers. And, moreover, the culture of such experiments, in fact very subtle, decayed and to a certain extent was lost.

It is also true that the last part of Prandtl's comment is very deep and instructive. But it remained dormant and was not cast into a proper mathematical theory for the following technical reason. In the early thirties, and even long before, the mathematical techniques which were needed here were in sufficiently good shape. However, they were considered as something of a mathematical monstrosity with no practical applications. Only several decades later was it recognized (see Chapters 3–5) that many physical phenomena needed these techniques for modelling, and then they entered the practice of applied mathematics and theoretical physics as incomplete similarity, fractals and renormalization groups. These concepts will be used in the present chapter to explain the situation regarding the scaling laws for turbulent shear flows at very large Reynolds numbers. In particular, incomplete similarity will allow a resolution of the contradiction mentioned in the last part of Prandtl's comment.

After von Kármán's (1930) work, Prandtl (1932) also arrived at the universal logarithmic law, using a different approach, and the term 'von Kármán–Prandtl universal logarithmic law' became established. Many different derivations of the universal logarithmic law were proposed later (e.g. Lighthill 1968, pp. 116–17; Landau and Lifshitz 1987, pp. 172–5; Schlichting 1968, pp. 489–90; Monin and Yaglom 1971, pp. 273–4; and, quite recently, Spurk 1997). We emphasize that the basis of all these derivations remained the hypothesis explicitly formulated by von Kármán, cited above. The only correction, proposed for the first time by Landau (see Landau and Lifshitz 1987), was that the hypothesis of viscosity independence was applied to the velocity gradient, not the velocity itself (see below).

The second major breakthrough in the theory of turbulence at very large Reynolds numbers happened in 1941 in the fundamental works of A.N. Kolmogorov and A.M. Obukhov, at that time Kolmogorov's student (Kolmogorov 1941; Obukhov 1941), where laws for the local structure of such flows were obtained. We emphasize particularly the role of the elucidating paper by Batchelor (1947), where the Kolmogorov–Obukhov theory, presented originally in the form of short notes, was explained in detail and fundamentally clarified. The problems of the local structure of developed turbulence are however outside the scope of the present book, mainly due to the lack of a sufficient experimental database, which is the only thing that can allow us to come to some ultimate conclusions.

8.2 Chorin's mathematical example

A.J. Chorin proposed a remarkable mathematical example, which elucidates the non-trivial mathematical situation in the problem of turbulent shear flows at large Reynolds numbers. Consider a family of curves

$$\phi = \left(\ln \frac{d}{\delta}\right)\left(\frac{y}{\delta}\right)^{1/\ln(d/\delta)} - 2\ln \frac{d}{\delta} \tag{8.3}$$

where ϕ is a dimensionless function, d and δ are parameters with the dimension of length and y is the independent variable, also having the dimension of length; $y > \delta$. We assume that d is fixed and that δ is the parameter of the family.

It is easy to show that the function ϕ satisfies the ordinary differential equation

$$\frac{d^2\phi}{dy^2} = \left(\frac{1}{\ln(d/\delta)} - 1\right)\frac{1}{y}\frac{d\phi}{dy} \tag{8.4}$$

and the boundary conditions

$$\phi(\delta) = -\ln \frac{d}{\delta}, \qquad \frac{d\phi}{dy}\bigg|_{y=\delta} = \frac{1}{\delta}. \tag{8.5}$$

Assume now that d is much larger than δ, $d \gg \delta$, so that $1/\ln(d/\delta)$ is a small parameter. For the curves of the family (8.3) a simple relation is easily obtained:

$$y\partial_y \phi = \left(\frac{y}{\delta}\right)^{1/\ln(d/\delta)} = \exp\left[\frac{\ln(y/d) + \ln(d/\delta)}{\ln(d/\delta)}\right]. \tag{8.6}$$

This relation shows that as $d/\delta \to \infty$ and for any fixed y/d the quantity $y\partial_y \phi = \partial_{\ln y}\phi$ tends to e.

As a function of δ the family (8.3) has an envelope

$$\phi = \ln\frac{y}{d}. \tag{8.7}$$

The quantity $\partial_{\ln y}\phi$ for the envelope is also a constant, but a different one, equal to unity. (We emphasize that here we consider only the branch of the family (8.3) having $d > \delta$. There is another branch with $d < \delta$, which also has an envelope $\phi = 2\ln[(y/d)(2-z)^{-1}]$, where $z = 1 \, 5936 \ldots$ is the second, non-zero, root of the equation $(2-z)\exp z = 2$. However, for our applications this branch is irrelevant because the basic length scale (for example the pipe diameter) is much larger than the viscous length scale δ, mentioned in the Prandtl's comment to the von Kármán lecture.)

Assume now that in equation (8.4), i.e. for $y > \delta$ but not in the boundary condition at $y = \delta$, we neglect (remember von Kármán's basic hypothesis!) the small parameter $1/\ln(d/\delta)$ in comparison with unity, so that equation (8.4) reduces to the form

$$\frac{d^2\phi}{dy^2} = -\frac{1}{y}\frac{d\phi}{dy}. \tag{8.8}$$

Satisfying the (δ-dependent!) boundary conditions (8.5) we obtain not the family (8.3) but only a single curve, the envelope (8.7), which is δ-independent ('universal'!). In fact, by neglecting the small parameter $1/\ln(d/\delta)$ in equation (8.4) we have prevented (cf. Prandtl's comment in the previous section) penetration of the influence of the parameter δ into the basic region.

Let us look at this matter from a different viewpoint. The derivative $d\phi/dy$ can be represented without solving equation (8.4) by dimensional analysis only, in the form

$$\frac{d\phi}{dy} = \frac{1}{y}\Phi\left(\frac{y}{\delta}, \frac{d}{\delta}\right) \tag{8.9}$$

where Φ is a dimensionless function of its dimensionless arguments. In the case under consideration,

$$\Phi = \left(\frac{y}{\delta}\right)^{1/\ln(d/\delta)}. \tag{8.10}$$

We see that at arbitrarily large y/δ the function Φ cannot be replaced by a constant, so that the influence of δ is preserved and cannot be neglected. However, δ enters the resulting equations in a specific, power-type form, due to a specific type of invariance of the problem as a whole. In fact, we have met in this example incomplete similarity in the parameter y/δ, explained in general terms in Chapter 4. We will see that the same situation happens in wall-bounded turbulent shear flows. However, if we do make the assumption of complete similarity, $\Phi =$ constant as above, then we recover the envelope of the family of solutions rather than the family of solutions themselves!

8.3 Steady shear flows at very large Reynolds numbers. The intermediate region in pipe flow[1]

We consider now the problem of statistically steady turbulent shear flows[2] ('hydraulic flows' in von Kármán terminology). Among such flows are many of practical importance, such as flows in pipes, channels and boundary layers. Their fundamental value is related also to their localised nature. In general, turbulent flows are non-local both in time and space, so that their mean properties are determined not only by the flow state at a given point but also by the flow history and the flow properties at neighboring points. This is not so for steady turbulent shear flows, and their localized nature radically simplifies our study of them. Flows in cylindrical pipes (Figure 8.1) constitute an instructive example of wall-bounded turbulent shear flows.

We have the same clear goal and well-determined problems as once formulated by von Kármán: to obtain mathematical expressions for the drag coefficient and the velocity distribution in the intermediate region of the flow. 'Intermediate' means outside the viscous sublayer, adjacent to the wall, where the velocity gradients are so high that the viscous stress is comparable with the stress created by turbulent vortices, and not too close to the pipe axis. Von Kármán also considered the same intermediate region of flow.

However, our basic hypothesis will be essentially different from von Kármán's hypothesis, presented in section 8.1, and this difference will lead to substantially different results. In fact we will replace von Kármán's hypothesis of complete similarity by the hypothesis of incomplete similarity.

---

[1] In the remaining part of this chapter the results of joint work of A.J. Chorin, V.M. Prostokishin and the author, performed over the period 1991–2000, are presented. Detailed references can be found in Barenblatt, Chorin and Prostokishin (1997, 2000a); Chorin (1998); Barenblatt (1999).

[2] Shear flows are flows with parallel mean velocities varying only in the lateral direction.

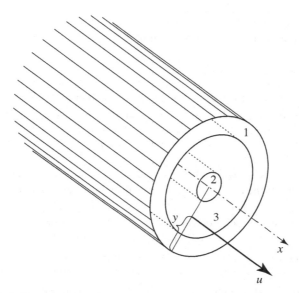

Figure 8.1. Flow in a long cylindrical pipe: the structure at large Reynolds number: 1, viscous sublayer; 2, near-axis region; 3, intermediate region.

We turn now to the derivation of the velocity distribution in the intermediate region. The mean velocity gradient $\partial_y u$ in the shear flow bounded by a smooth wall depends on the following arguments: the transverse coordinate y (the distance from the wall), the shear stress at the wall τ, the pipe diameter d and the fluid properties, its kinematic viscosity ν and density ρ. *The velocity gradient $\partial_y u$ is considered rather than the velocity u itself, because the values of u at an arbitrary distance from the wall depend on the flow in the vicinity of the wall, where intrinsically asymptotic assumptions, which we will use below, are clearly invalid.* Thus

$$\partial_y u = f(y, \tau, d, \nu, \rho). \tag{8.11}$$

Following von Kármán and Prandtl we introduce the viscous length scale

$$\delta = \frac{\nu}{u_*}, \qquad \text{where} \quad u_* = \sqrt{\frac{\tau}{\rho}} \tag{8.12}$$

(the quantity u_* is called the dynamic or frictional velocity), and a standard application of dimensional analysis then gives

$$\partial_y u = \frac{u_*}{y} \Phi\left(\frac{y}{\delta}, \frac{d}{\delta}\right). \tag{8.13}$$

Also, dimensional analysis of the relation (8.11) shows that $d/\delta = u_* d/\nu$ is a function of the traditional Reynolds number,

$$Re = \frac{\bar{u}d}{\nu}, \tag{8.14}$$

where $\bar{u}$ is the average velocity – the flux divided by the cross-sectional area of the pipe. The relation (8.13) can be rewritten therefore as

$$\partial_y u = \frac{u_*}{y} \Phi \left(\frac{y}{\delta}, Re \right). \tag{8.15}$$

For very large Reynolds numbers in the intermediate region under consideration, the ratio of the distance from the wall to the viscous length scale y/δ is large. The basic von Kármán hypothesis (see section 8.1) is that the viscosity does not affect the velocity distribution in this region. However, in the expression (8.15) for the velocity gradient the viscosity enters both arguments, y/δ and Re. Therefore this hypothesis means complete similarity in the parameters y/δ and Re. More explicitly it means that the function Φ tends to a finite non-zero limit as its arguments tend to infinity independently. According to von Kármán's hypothesis the viscous length scale δ should disappear from the resulting relations, and the function Φ can be replaced by a constant: $\Phi = 1/\kappa$. The constant κ was later named 'Kármán's constant'. Substituting $\Phi = 1/\kappa$ into (8.15) gives

$$\partial_y u = \frac{u_*}{\kappa y}. \tag{8.16}$$

Integration gives the von Kármán–Prandtl universal (Reynolds-number independent) logarithmic law for the velocity distribution:

$$u = u_* \left(\frac{1}{\kappa} \ln \frac{u_* y}{\nu} + C \right), \tag{8.17}$$

where the constant C is finite and Re-independent (and this also is a seemingly logically consistent, but nevertheless a substantial extra, assumption). Prandtl (see his comment in section 8.1) emphasized a 'seeming contradiction' related to the appearance of the viscosity in the resulting formula, which will be explained later.

For more than six decades the experimental information accumulated, suggested some doubts in the universal logarithmic law, i.e. in the von Kármán basic hypothesis, which we now call the hypothesis of complete similarity. The experimental data demonstrates a systematic deviation (not a scatter!) from the predictions of the universal logarithmic law even if a very liberal approach to the constants κ and C is allowed (κ from 0.38 to 0.44, i.e. $1/\kappa$ from 2.25 to 2.65; C from 4.1 to 6.3!), although by the very logic of their derivation these constant

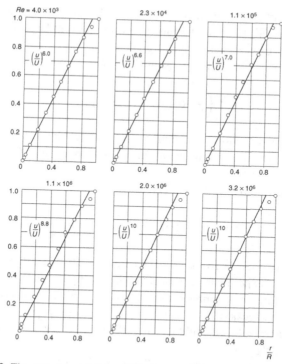

Figure 8.2. The experimental data of Nikuradze processed in the form of power laws $(u/U)^{n(Re)}$ versus r/R show that the power laws give a good approximation for the velocity distribution data. Here U is the maximum velocity, r the distance from the wall (Schlichting (1968)). The variation of n is significant: from $1/6.0$ for $Re = 4.0 \cdot 10^3$ to $1/10$ for $Re = 3.2 \cdot 10^6$

should be identical for all high-quality experiments in smooth pipes. Moreover, the processing of the experimental data of Nikuradze, mentioned by Prandtl in his comments to the von Kármán lecture as a good confirmation of Kármán's formula, suggested (see Figure 8.2) that a power law with the power depending on Re could be a good representation of the experimental data concerning the velocity distribution. Therefore it was a natural step to assume that there is no complete similarity and to propose instead of the von Kármán hypothesis a different hypothesis, suggesting the step next in complexity:

First hypothesis: *There is an incomplete similarity of the average velocity gradient in the parameter y/δ and no kind of similarity in Re.*

According to this hypothesis, the influence of the viscosity remains at arbitrary large Reynolds numbers in the whole body of the flow, but it enters only in combination with other parameters controlling the turbulence. Practically, this means that for very large Re the function Φ in (8.15) at large y/δ should be assumed to be a power function of its argument y/δ, while no special suggestion of any kind of similarity in Re is assumed, so that

$$\Phi\left(\frac{y}{\delta}, Re\right) = A(Re)\left(\frac{y}{\delta}\right)^{\alpha(Re)} \tag{8.18}$$

where $A(Re)$ and $\alpha(Re)$ are certain, as yet undetermined, functions of the Reynolds number. It is instructive at this point to remember Chorin's example presented in section 8.2.

Substituting (8.18) in (8.15) we obtain

$$\partial_y u = \frac{u_*}{y} A(Re)\left(\frac{y}{\delta}\right)^{\alpha(Re)}, \qquad \delta = \frac{\nu}{u_*}. \tag{8.19}$$

Note that the relation (8.16) is a special case of (8.19). Therefore, the original idea was that if experiments and/or numerical computations (for the relevant range of high Reynolds numbers, numerical computations are at present impossible, so here we speak of the rather distant future) showed that A is a universal constant while $\alpha = 0$, we could return to (8.16). *Now we can claim definitely that this is not the case!*

Note immediately a clear-cut qualitative difference between the cases of complete and incomplete similarity. In the first case, the experimental data should cluster in the traditional $(\ln \eta)\phi$-plane ($\phi = u/u_*$, $\eta = u_* y/\nu = y/\delta$) on the single straight line of the universal logarithmic law. In the second case, the experimental points should occupy an area in the $(\ln \eta)\phi$-plane; a separate curve corresponds to each value of the Reynolds number.

Our next hypothesis will be the *vanishing-viscosity principle:*

Second hypothesis: *The gradient of the average velocity tends to a well-defined limit as the viscosity vanishes.*

This principle is in clear correspondence with the last part of Prandtl's conclusion, and was also used implicitly by von Kármán (see section 8.1).

Experiments even at high Reynolds numbers demonstrate a perceptible dependence of the dimensionless velocity distribution $\phi = u/u_*$ on Re (see Figure 8.3). Therefore, and according to the vanishing viscosity principle, it is appropriate to expand $A(Re)$ and $\alpha(Re)$ in a series in a small parameter $\varepsilon(Re)$ that vanishes at $Re = \infty$, and to retain the first two terms:

$$A = A_0 + A_1\varepsilon, \qquad \alpha = \alpha_0 + \alpha_1\varepsilon$$

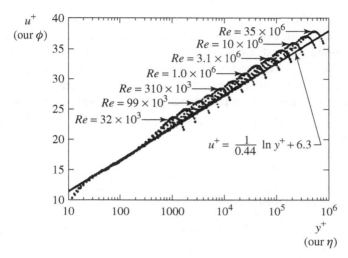

Figure 8.3. The Princeton data (Zagarola 1996) obtained in a high-pressure pipe confirm the splitting of the experimental data according to their Reynolds numbers and the rise of the curves above their envelope in the $(\ln \eta)\phi$-plane $(\eta = u_*y/\nu, \phi = u/u_*.)$ The solid line is the envelope; each dotted curve has its maximum at the η-value corresponding to the center of the pipe. The splitting and form of the curves agree with the scaling law and are incompatible with the von Kármán–Prandtl universal logarithmic law.

where A_0, A_1, α_0 and α_1 should be, by the logic of the derivation, universal Reynolds-number-independent constants. We obtain from (8.19)

$$\partial_y u = \frac{u_*}{y}(A_0 + A_1 \varepsilon)\left(\frac{y}{\delta}\right)^{\alpha_0 + \alpha_1 \varepsilon}. \qquad (8.20)$$

When the viscosity (and, consequently, the length scale δ) tends to zero, a well-defined limit of (8.20) exists only for $\alpha_0 = 0$; therefore, according to our second hypothesis $\alpha_0 = 0$. There is also the possibility that $A_0 = 0$, $\alpha_0 \neq 0$ and $\varepsilon = (Re)^{-\alpha_0}$; it leads to a universal power dependence of $\partial_y u$ upon y. This possibility, however, is found not to be compatible with the experimental data (cf. Figure 8.2).

Furthermore, (8.20) can be represented as

$$\partial_y u = \frac{u_*}{y}(A_0 + A_1 \varepsilon) \exp\left(\alpha_1 \varepsilon \ln \frac{y}{\delta}\right)$$
$$= \frac{u_*}{y}(A_0 + A_1 \varepsilon) \exp\left(\alpha_1 \varepsilon \ln \frac{u_* d}{\nu} + \alpha_1 \varepsilon \ln \frac{y}{d}\right). \qquad (8.21)$$

The small parameter ε is a function of the Reynolds number vanishing at $Re = \infty$. Relation (8.21) shows that if ε tended to zero as $Re \rightarrow \infty$ faster than $1/\ln Re$ then the argument of the exponent would tend to zero and we would return to the case of complete similarity. The experiments, as was mentioned before, show that this is not the case. If ε tends to zero slower than $1/\ln Re$, however, a well-defined limit of the velocity gradient as the viscosity goes to zero does not exist, and we obtain a contradiction to the second hypothesis, the vanishing-viscosity principle. *Therefore, the only choice compatible with our basis hypotheses (incomplete similarity and the vanishing-viscosity principle)* is

$$\varepsilon = \frac{1}{\ln Re}. \tag{8.22}$$

Thus we obtain, by integration of (8.21),

$$\phi = \frac{u}{u_*} = (C_0 \ln Re + C_1)\left(\frac{y}{\delta}\right)^{\alpha_1/\ln Re}. \tag{8.23}$$

Here an additional condition

$$\phi(0) = 0$$

has been used. This condition is an independent assumption, confirmed by experiment, which does not follow from the non-slip boundary condition $u(0) = 0$, because the boundary $y = 0$ is outside the range of applicability of the intermediate-asymptotic relation (8.21).

Thus, we arrive at a conclusion which is in correspondence with the intuitive idea of Prandtl (see the last sentence of his comment). Indeed, *the wall streaks where the turbulence and viscosity act together penetrate to the main body of the flow at any finite Reynolds numbers.* And it is clear that these same streaks create the *intermittency* of wall-bounded flows. This conclusion is also in correspondence with the idea of incomplete similarity.

The parameters of turbulence (u_*, Re) and viscosity (ν) form together a monomial

$$C = (C_0 \ln Re + C_1)u_*^{1+\alpha_1/\ln Re}\nu^{-\alpha_1/\ln Re}, \tag{8.24}$$

whose dimension cannot be obtained from dimensional analysis and which determines the velocity distribution

$$u = C\, y^{\alpha_1/\ln Re}.$$

A careful comparison with the data of Nikuradze's experiments, which were performed under the direct guidance of Prandtl, suggested (see the details and further references in Barenblatt, Chorin and Prostokishin 1997) the following values of the universal constants:

$$C_0 = \frac{1}{\sqrt{3}}, \qquad C_1 = \frac{5}{2}, \qquad \alpha_1 = \frac{3}{2}. \qquad (8.25)$$

Therefore *the ultimate scaling law proposed for the velocity distribution in the major, intermediate, region of the pipe is*

$$\phi = \left(\frac{\sqrt{3} + 5\alpha}{2\alpha} \right) \eta^\alpha, \qquad \alpha = \frac{3}{2 \ln Re}, \qquad \phi = \frac{u}{u_*}, \qquad \eta = \frac{u_* y}{\nu}. \qquad (8.26)$$

The scaling law (8.26) shows that, as expected, there is no universal Re-independent velocity distribution in the $(\ln \eta)\phi$-plane but that there is a family of curves in this plane with Re as a parameter. However, the family (8.26) has the special property of self-similarity and therefore of universality. Indeed, if we plot on the ordinate axis instead of ϕ the quantity

$$\psi = \frac{1}{\alpha} \ln \frac{2\alpha\phi}{\sqrt{3} + 5\alpha}, \qquad \alpha = \frac{3}{2 \ln Re} \qquad (8.27)$$

we obtain $\psi = \ln \eta$, i.e. the equation of the bisectrix of the first quadrant. Comparison with Nikuradze's experimental data shows that this is the case (Figure 8.4): the overwhelming majority of the experimental points for $\eta > 30$ do indeed settle down to the bisectrix. The points corresponding to $\eta < 30$ naturally deviate from the bisectrix, because the scaling law (8.26) describes the velocity distribution in the intermediate part of the cross-section, but it should be emphasized that this deviation is a systematic one, not a scatter.

The scaling law (8.26) allows determination of the dependence of the drag coefficient on the Reynolds number. We define the dimensionless skin-friction drag coefficient in a way now common in the literature:

$$\lambda = \frac{\tau}{\rho \bar{u}^2 / 8} = 8 \left(\frac{u_*^2}{\bar{u}^2} \right). \qquad (8.28)$$

Using for the determination of the average velocity $\bar{u}$ the scaling law (8.26), and neglecting the deviation of the velocity distribution from the scaling law in the viscous sublayer and near the axis, we obtain the formula for the *skin-friction coefficient as a function of the Reynolds number*:

$$\lambda = \frac{8}{\Psi^{2/(1+\alpha)}}, \qquad \Psi = \frac{e^{3/2}(\sqrt{3} + 5\alpha)}{2^\alpha \alpha (1 + \alpha)(2 + \alpha)}, \qquad \alpha = \frac{3}{2 \ln Re}. \qquad (8.29)$$

Comparison of this law with Nikuradze's independent series of experiments (Nikuradze 1932) determining the skin friction also showed an instructive

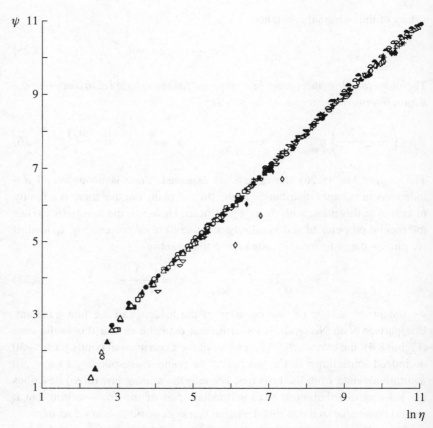

Figure 8.4. The experimental data of Nikuradze (1932) in the coordinates $\ln \eta$, ψ at $\eta > 30$ lie close to the bisectrix of the first quadrant, confirming the scaling law. The values of Re are as follows: $\triangle$, 4×10^3; $\blacktriangle$, 6.1×10^3; $\circ$, 9.1×10^3; $\bullet$, 1.67×10^4; $\square$, 2.33×10^4; $\blacksquare$, 4.34×10^4; $\triangledown$, 1.05×10^5; $\blacktriangledown$, 2.05×10^5; $\circ$, 3.96×10^5; $\bullet$, 7.25×10^5; $\diamond$, 1.11×10^6; $\blacklozenge$, 1.536×10^6; $+$, 1.959×10^6; $\times$, 2.356×10^6; $\circ$, 2.79×10^6; $\blacktriangle$, 3.24×10^6.

agreement (see further details and references in Barenblatt, Chorin and Prostokishin 1997). The deviations are within the limits of a normal experimental scatter.

We arrive at the conclusion that the scaling law (8.26) with universal constants (8.25) and the drag law (8.29) describe the flow in smooth pipes satisfactorily for large Reynolds numbers and that the incomplete similarity of this flow can be considered as established. Professor N. Zabusky proposed the very appropriate term 'experimental asymptotics' for the style of argument used here for the derivation of the scaling law (8.26).

8.4 Modification of Izakson–Millikan–von Mises derivation of the velocity distribution in the intermediate region. The vanishing-viscosity asymptotics

The universal logarithmic law hardened into dogma, and became one of the pillars of turbulence theory and a mainstay of engineering science, to a large extent because it was supported by an independent mathematical derivation based on seemingly unassailable principles. This derivation was proposed by Izakson (1937), Millikan (1939) and von Mises (1941); see also Monin and Yaglom (1971), pp. 299–301. It is usually presented as follows. It is assumed that in the intermediate region under consideration the dimensionless velocity distribution is a universal, Reynolds-number-independent, function of the local Reynolds number $\eta = u_* y / \nu$. Thus, the influence of the external dimensional parameter, the pipe diameter d, and consequently that of the Reynolds number, is neglected, so that on the one hand the *wall law* is valid:

$$\phi = \frac{u}{u_*} = f\left(\frac{u_* y}{\nu}\right),$$ (8.30)

where f is a certain dimensionless function.

On the other hand, in the vicinity of the pipe axis the *defect law* is assumed to be valid:

$$u_{\mathrm{CL}} - u = u_* g\left(\frac{2y}{d}\right)$$ (8.31)

where u_{CL} is the mean velocity at the pipe axis, so that $u_{\mathrm{CL}} - u$ is the velocity defect, and g is another dimensionless function, but of a different argument; in relation (8.31) the influence of viscosity is neglected. Thus it is assumed that in the wall region the influence of the external length scale d can be neglected, whereas near the pipe axis the influence of the internal length scale $\delta = \nu/u_*$ can be neglected. The next step is the assumption that there exists at very large Reynolds numbers an interval of distance where both laws (8.30) and (8.31) are valid. Therefore a functional equation

$$u_{\mathrm{CL}} - u_* f\left(\frac{u_* y}{\nu}\right) = u_* g\left(\frac{2y}{d}\right)$$ (8.32)

is obtained by combining (8.30) and (8.31). After differentiation of (8.32) by y followed by multiplication by y the following relation is obtained:

$$\eta f'(\eta) = -\xi g'(\xi);$$ (8.33)

here $\xi = 2y/d$. The right- and left-hand sides of equation (8.33) contain functions of different arguments, therefore each of the sides can be only a constant.

Denoting this constant by $1/\kappa$ and integrating, the 'law of the wall' is obtained in the form of the universal logarithmic law

$$f(\eta) = \frac{1}{\kappa} \ln \eta + B; \tag{8.34}$$

the defect law

$$g(\xi) = -\frac{1}{\kappa} \ln \xi + B_0 \tag{8.35}$$

is also obtained, with

$$B_0 = \frac{u_{\mathrm{CL}}}{u_*} - \frac{1}{\kappa} \ln \frac{u_* d}{2\nu} - B. \tag{8.36}$$

This very attractive derivation was apparently one of the first applications of the method of matched asymptotic expansions, which is very popular nowadays; see the illuminating monographs Van Dyke (1975), Cole (1968) and Hinch (1991).

The derivation is however not quite correct and needs a modification. It is clear now (cf. Figure 8.3) that neither in the law of the wall nor in the defect law can the influence of the Reynolds number be neglected. Indeed, there exist rectilinear parts of the curves in Figure 8.3, but they are different for different Reynolds numbers. Furthermore, the domes of the curves corresponding to the velocity maximum are also different for different Reynolds numbers. So, according to our basic concept, these laws should be represented in the form

$$\phi = \frac{u}{u_*} = f\left(\frac{u_* y}{\nu}, Re\right) \tag{8.37}$$

and

$$u_{\mathrm{CL}} - u = u_* g\left(\frac{2y}{d}, Re\right). \tag{8.38}$$

The derivation thereafter proceeds as before, the only differences being that κ is no longer a constant but, rather, a certain function of the Reynolds number, $\kappa = \kappa(Re)$, and that B, – this is very significant, – is also a function of the Reynolds number.

Therefore the law of the wall assumes the form

$$\phi = \frac{u}{u_*} = \frac{1}{\kappa(Re)} \ln \frac{u_* y}{\nu} + B(Re), \tag{8.39}$$

where $\kappa(Re)$ and $B(Re)$ are certain unspecified functions.

It is essential that there is no contradiction between the scaling law (6.4) and the law of the wall (8.39). This was demonstrated in Barenblatt and Chorin (1996, 1997), where the vanishing-viscosity method developed by Chorin (see

Chorin 1988, 1994) was used essentially. Indeed, the law (8.26) can be written in the form

$$\phi = \left(\frac{1}{\sqrt{3}} \ln Re + \frac{5}{2} \right) \exp \left(\frac{3 \ln \eta}{2 \ln Re} \right). \tag{8.40}$$

Let the observation point be at a fixed distance y from the wall that is definitely larger than a certain length Δ, for instance, the size of a gauge. Let the pipe diameter and the pressure gradient be fixed also. One is not free to vary $Re = \bar{u}d/\nu$ and $\eta = u_* y/\nu$ independently because the viscosity ν appears in both. So, if ν is decreased by the experimenter, as it was in the Princeton Super-pipe experiments (Zagarola *et al.* 1996; Zagarola 1996), whose basic idea was proposed by Brown (1991), one considers flows at ever larger η and ever larger Re; in particular the lowest $\eta = u_* \Delta/\nu$ increases with decreasing viscosity. Consider now the ratio $3 \ln \eta/(2 \ln Re)$, which enters the scaling law in the form seen in (8.40). We have from the definitions of η and Re

$$\frac{3 \ln \eta}{2 \ln Re} = \frac{3}{2} \left[\ln \frac{\bar{u}d}{\nu} + \ln \frac{y}{d} + \ln \frac{u_*}{\bar{u}} \right] \frac{1}{\ln(\bar{u}d/\nu)}. \tag{8.41}$$

However, the distance from the wall y lies in the fixed interval $\Delta < y < d/2$, and $\bar{u}/u_*$ can be shown to be of order $\ln Re$, so that $\ln(u_*/\bar{u}) \sim \ln \ln Re$, which is asymptotically small at very large Re. Therefore $3 \ln \eta/(2 \ln Re)$ is asymptotically close to 3/2, and the quantity

$$1 - \frac{\ln \eta}{\ln Re}$$

can be considered to be a small parameter, so that

$$\exp \left[\frac{3}{2} \frac{\ln \eta}{\ln Re} \right] \approx \exp \left[\frac{3}{2} - \frac{3}{2} \left(1 - \frac{\ln \eta}{\ln Re} \right) \right]$$

$$= e^{3/2} \left[1 - \frac{3}{2} \left(1 - \frac{\ln \eta}{\ln Re} \right) \right]$$

$$= e^{3/2} \left[\frac{3}{2} \frac{\ln \eta}{\ln Re} - \frac{1}{2} \right]. \tag{8.42}$$

This means that in the interval of interest, $\Delta < y < d/2$, the power law (8.26) can be approximated by

$$\phi = e^{3/2} \left(\frac{\sqrt{3}}{2} + \frac{15}{4 \ln Re} \right) \ln \eta - \frac{e^{3/2}}{2\sqrt{3}} \ln Re - \frac{5}{4} e^{3/2} \tag{8.43}$$

i.e. by a relation of the form (8.39), with

$$\kappa(Re) = \frac{e^{-3/2}}{\sqrt{3}/2 + 15/(4 \ln Re)}, \qquad B(Re) = -\frac{e^{3/2} \ln Re}{2\sqrt{3}} - \frac{5}{4} e^{3/2}. \quad (8.44)$$

It is important that as $Re \to \infty$ the value of $\kappa(Re)$ tends to a finite non-zero limit, $2/(\sqrt{3}e^{3/2}) \simeq 0.2776$, whereas the additive constant B, which has no finite limit, tends to $-\infty$.

At the same time, the family of power laws (8.26) having Re as parameter has an envelope (cf. Chorin's example presented in section 8.2). The relation for the envelope is obtained in implicit form by the elimination of Re between equation (8.26) and the equation

$$\frac{3 \ln \eta}{2 \ln Re} = \frac{\sqrt{3}}{10} \ln \eta \left[\left(1 + \frac{20}{\sqrt{3} \ln \eta} \right)^{1/2} - 1 \right], \quad (8.45)$$

which is obtained from (8.26) by differentiation with respect to Re. And the envelope has an important feature: in the working range of $\ln \eta$ it is practically indistinguishable from the straight line

$$\phi = \frac{u}{u_*} = \frac{\sqrt{3}e}{2} \ln \eta + 5.1 \quad (8.46)$$

(see Barenblatt 1993; Barenblatt and Prostokishin 1993; Barenblatt, Chorin and Prostokishin 1997). Bearing in mind that $2/(\sqrt{3}e) = 0.425$, the straight line (8.46) can be identified with the traditional form of the universal logarithmic law. Therefore if one plots the experimental points that correspond to various values of Re, and to y-values that are not too large, on a single graph in the $(\ln \eta)\phi$-plane what is natural for those who happen to believe in the universal logarithmic law is that the envelope will be revealed. The visual impact of the envelope, when plotting the experimental data in the $(\ln \eta)\phi$-plane, is magnified by the fact that measurements at very small values of y, where the difference between the power laws and the envelope can also be noticeable, is missing due to experimental difficulties. Thus if the proposed scaling law (8.26) is valid, the seeming confirmation of the universal logarithmic law is nothing but an illusion. The characteristic features of the Reynolds-number-dependent scaling law, in addition to the splitting of the curves according to their Reynolds number, are the presence of straight-line parts at very large Reynolds numbers and a discrepancy of about $\sqrt{e}$ between the slopes of the curves and the slope of the envelope (Figure 8.5). These qualitative distinctions are confirmed by the experiments of the Princeton group (see Figure 8.2). Indeed, despite a flaw in these experiments, discussed in detail in Barenblatt, Chorin and Prostokishin

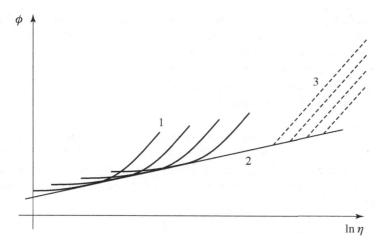

Figure 8.5. A schematic diagram of the power-law curves in a pipe, their envelope and their asymptotic slope. The apparent motion of the curves to the right corresponds to the changes in Reynolds number. 1, the velocity as a function of the distance from the wall (in appropriate units); 2, the envelope of the power laws, formerly mistaken for the curves themselves; 3, the asymptotic rectilinear parts of the law-of-the-wall curves.

(1997) and in Perry *et al.* (2001), the results are sufficiently robust to exhibit a separate curve for each Reynolds number and a well-defined angle between the rectilinear parts of the curves and their envelope.

8.5 Turbulent boundary layers

The universal law for pipe flow (8.16) can be represented in the dimensionless form

$$\eta \partial_\eta \phi = \frac{1}{\kappa} \tag{8.47}$$

and the Reynolds-number-dependent scaling law (8.26) in a corresponding form,

$$\eta \partial_\eta \phi = \left(\frac{\sqrt{3}}{2} + \frac{15}{4 \ln Re} \right) \eta^{3/(2 \ln Re)}. \tag{8.48}$$

The laws (8.17), (8.26) for the dimensionless velocity ϕ may be obtained from (8.47) and (8.48) integration.

By the logic of its derivation, the scaling law (8.48) must be valid not only for the flow in pipes but also for any turbulent wall-bounded shear flows.

Here, however, a basic question appears – what is the definition of the Reynolds number for these flows which would allow the use of the law (8.26) for them? This basic question is immaterial as long as the engineer or researcher continues to believe in the universal logarithmic law. Indeed, if the law is Re-independent, the definition of Re does not matter.[3] The situation is different though, when the law is Re-dependent. But what should one do for other shear flows?

We will consider below boundary layers and will show that the law (8.26) also describes these flows under a appropriate choice of Reynolds numbers. Zero pressure-gradient boundary layers have been well investigated experimentally over the last 25 years. The common choice of Reynolds numbers for these flows is

$$Re_\theta = \frac{U\theta}{\nu}, \qquad \theta = \frac{1}{U^2} \int_0^\infty u(U - u)\, dy \qquad (8.49)$$

where U is the free stream velocity and θ is a length calculated by integration of the velocity profile, the so-called momentum thickness. This choice is rather arbitrary, and a priori the law (8.26) with $Re = Re_\theta$ should not be valid. But what is the proper choice of Re for the boundary layers?

To understand this, we have to confirm first of all that in the intermediate layer of the boundary layer flow adjacent to the viscous sublayer *a certain scaling law is valid*. To do that (see details in Barenblatt, Chorin, Hald and Prostokishin 1997 and in Barenblatt, Chorin and Prostokishin 2000a) all available experimental data presented in the traditional $(\ln \eta)\phi$-plane were replotted in a bilogarithmic $(\log_{10} \eta)\, \log_{10} \phi$-plane. The result was instructive: without exception, for all investigated flows a straight line was obtained for region I, the region adjacent to the viscous sublayer (see the examples in Figure 8.6). Moreover, for flows with low free-stream turbulence a second self-similar region, II, was observed between the first region and the free-stream flow. The analysis of this region II was performed in a recent paper by Barenblatt, Chorin and Prostokishin (2002). Its degradation and subsequent disappearance with growing free-stream turbulence was proved by the processing of the experiments of Hancock and Bradshaw (1989); see Barenblatt, Chorin and Prostokishin (2000a).

The straight line in region I corresponds to the scaling law

$$\phi = A\eta^\alpha; \qquad (8.50)$$

the coefficients A and α were obtained by statistical processing.

We assume that the effective Reynolds number Re has the form $Re = U\Lambda/\nu$ where, we repeat, U is the free-stream velocity and Λ is a certain length scale

---

[3] As Ya.B. Zeldovich used to joke: If you ask a barman to serve water without syrup, the question 'without which syrup' is inappropriate. But if the water is assumed to be with syrup, then the question 'which syrup should be served' is clearly relevant.

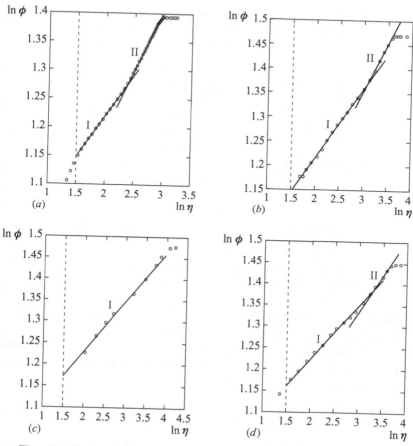

Figure 8.6. Data, replotted in bilogarithmic form, obtained from experiments turbulent boundary layers. (a) The data of Erm and Joubert (1991); $Re_\theta = 2788$. Both self-similar intermediate regions, I and II, are clearly seen. (b) The data of Krogstad and Antonia (1999); $Re_\theta = 12\,570$. Again, both self-similar intermediate regions, I and II, are clearly seen. (c) The data of Petrie, Fontaine, Sommer and Brugart obtained by scanning the graphs in Fernholz and Finley (1996); $Re_\theta = 35\,530$. The first self-similar region, I, is seen but the second self-similar region, II, is barely revealed. (d) The data of Smith, obtained by scanning the graphs in Fernholz and Finley (1996); $Re_\theta = 12\,990$. The first self-similar region, I, and the second region, II, are clearly seen.

The basic question is whether such a unique length scale Λ, which plays the same role for the intermediate region I of the boundary layer as does the diameter for the pipe flow, exists? In other words, is it possible to find a length scale Λ, perhaps influenced by individual features of the flow, such that the scaling law (8.26) is valid for the first intermediate region I? To answer this question, in

Table 8.1

Re_θ	α	A	$\ln Re_1$	$\ln Re_2$	$\ln Re$	Re_θ/Re
Erm and Joubert (1991)						
697	0.163	7.83	9.23	9.20	9.22	0.07
1003	0.159	7.96	9.46	9.43	9.45	0.08
1568	0.156	7.99	9.51	9.62	9.56	0.11
2226	0.148	8.28	10.01	10.14	10.07	0.09
2788	0.140	8.66	10.67	10.71	10.69	0.06
Krogstad and Antonia (1999)						
12 570	0.146	8.38	10.18	10.27	10.23	0.45
Petrie, Fontaine, Sommer and Brungart[a]						
35 530	0.119	9.76	12.57	12.61	12.59	0.12
Smith[a]						
4996	0.146	8.36	10.15	10.27	10.21	0.18
12 990	0.130	9.08	11.40	11.54	11.47	0.14

[a] The data were obtained by scanning the graphs in the review Fernholz and Finley (1996).

Barenblatt, Chorin, Hald and Prostokishin (1997) and in Barenblatt, Chorin and Prostokishin (2000a) A and α (obtained, we emphasize, by statistical processing of the experimental data in the first intermediate scaling region) were taken and two values, $\ln Re_1$ and $\ln Re_2$, were then calculated by solving the two equations suggested by the scaling law (8.26),

$$\frac{1}{\sqrt{3}} \ln Re_1 + \frac{5}{2} = A, \qquad \frac{3}{2 \ln Re_2} = \alpha. \qquad (8.51)$$

If these values of $\ln Re_1$ and $\ln Re_2$ obtained by solving two different equations, are indeed close, i.e. if they coincide to within experimental accuracy, it means that the unique length scale Λ can be determined and the experimental scaling law in region I coincides with the basic scaling law (8.26).

Table 8.1 shows via several examples that the values of $\ln Re_1$ and $\ln Re_2$ are close; a more detailed discussion of all the available data can be found in Barenblatt, Chorin and Prostokishin (2000a), but the conclusion remains the same. Owing to the closeness of the values of $\ln Re_1$ and $\ln Re_2$, we can introduce for all these flows a mean Reynolds number Re, for instance as

$$Re = \sqrt{Re_1 Re_2}, \qquad \ln Re = \frac{1}{2}(\ln Re_1 + \ln Re_2) \qquad (8.52)$$

and consider Re as an estimate for the effective Reynolds number of the boundary-layer flow. Naturally the ratio $Re_\theta/Re = \theta/\Lambda$ is different for different flows.

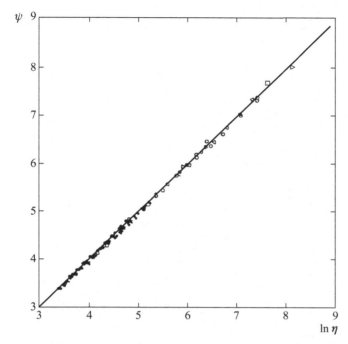

Figure 8.7. The data of Erm and Joubert (1991) (∗), Krogstad and Antonia (1999) (◁), Smith (□) and Petrie *et al.* (▷) collapse on the bisectrix of the first quadrant in the (ln η)ψ-plane, in accordance with the universal form (8.53) of the scaling law (8.26).

Checking the universal form of the scaling law (8.26),

$$\psi = \frac{1}{\alpha} \ln \left(\frac{2\alpha\phi}{\sqrt{3} + 5\alpha} \right) = \ln \eta, \tag{8.53}$$

gives another way to demonstrate clearly its applicability to the first intermediate region of the flow adjacent to the viscous sublayer. According to the relation (8.53), in the coordinates ln η, ψ all experimental points should collapse onto the bisectrix of the first quadrant. Seen in Figure 8.7 are the data of Erm and Joubert (1991), Krogstad and Antonia (1999), Smith and Petrie *et al.* (the data in the last two references were obtained by scanning the graphs in the review Fernholz and Finley 1996). It is seen that the data collapse onto the bisectrix with sufficient accuracy to confirm the scaling law (8.26). The parameter α was calculated according to the formula $\alpha = 3/(2 \ln Re)$, $\ln Re$ being taken as $(\ln Re_1 + \ln Re_2)/2$; see Table 8.1. The results of similar processing of all the available data can be found in Barenblatt, Chorin and Prostokishin (2000a).

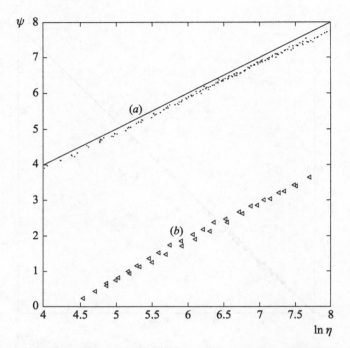

Figure 8.8. (a) The data of Naguib (1992) and of Nagib and Hites (1995) show a systematic deviation from the bisectrix of the first quadrant (straight line). (b) The data of Krogstad and Antonia (1998) relate to rough walls; the experimental points lie much lower than the bisectrix. For the evaluation of ψ the value $\alpha = 3/(2 \ln \mathrm{Re}_1)$ was taken.

We conclude that the scaling law (8.26) gives an accurate description of the mean velocity distribution over the self-similar intermediate region (I) adjacent to the viscous sublayer for a wide variety of zero-pressure-gradient boundary-layer flows. In the paper by Panton (2002) it is stated that '... the method that Barenblatt, Chorin and Prostokishin proposed to extend the power law to boundary layers displays extreme sensitivity.' Indeed, the ability of this method to reveal even small deviations in experimental data is illustrated in Figure 8.8. When the universality of the data of Naguib (1992) and Nagib and Hites (1995), obtained at the Illinois Institute of Technology (IIT), was checked, a small but systematic parallel shift from the bisectrix was revealed (Figure 8.8(a)). It would appear that the following sentence from the paper Österlund et al. (1999) explaining the details of the experiment at IIT can clarify the reason for this shift: 'A short fetch of sandpaper roughness was used to trigger the transition in the boundary layer at the same location for all similar velocities.' For comparison, in Figure 8.8(b) we give a presentation of the results of Krogstad and Antonia (1999) relating to rough walls; the parallel shift there is much larger.

The Reynolds number is defined as $Re = U \Lambda / \nu$, where U is the free-stream velocity and Λ is a length scale which is well defined for all the flows under investigation. An analysis of experimental data performed in Barenblatt, Chorin and Prostokishin (2000b) showed that Λ is between 1.5 and 1.6 times the wall-region thickness as determined by the sharp intersection of the two velocity distribution laws I and II. The validity of the scaling law (8.26) for the lower self-similar region, I, of the boundary-layer flows constitutes a strong argument in favor of its validity for a wide class of wall-bounded turbulent shear flows at large Reynolds numbers.

The nature of the second self-similar region, II, adjacent to the free stream, is not yet completely clear. For zero-pressure-gradient boundary layer flows in the absence of free-stream turbulence the power β in the scaling law valid for this region,

$$\phi = B\eta^{\beta}, \tag{8.54}$$

is close to 0.2. The data for non-zero-pressure-gradient boundary layers are substantially less numerous. Recently on the Internet the data of Marušić and Perry (1995) appeared in digital form. The processing of these data in Barenblatt, Chorin and Prostokishin (2002) confirmed the chevron-like structure of the velocity distribution of the boundary layer. It showed that the power β has a substantial variation (see Figure 8.9). Let us determine the set of parameters which govern the coefficient B and the power β in the scaling law (8.54). One of these parameters must be the dimensionless effective Reynolds number Re which determines the flow structure in layer I and is affected, in turn, by the flow in the viscous sublayer and in layer II. The following dimensional parameters should also influence the flow in the upper layer: the pressure gradient $\partial_x p$, the dynamic (friction) velocity u_* and the fluid properties, its kinematic viscosity ν and the density ρ. The dimensions of the governing parameters in the LMT class are as follows:

$$[\partial_x p] = \frac{M}{L^2 T^2}, \qquad [u_*] = \frac{L}{T}, \qquad [\nu] = \frac{L^2}{T}, \qquad [\rho] = \frac{M}{L^3}.$$

The first three parameters have independent dimensions. Therefore only one additional dimensionless governing parameter can be formed from the dimensional parameters:

$$P = \frac{\nu \partial_x p}{\rho u_*^3}. \tag{8.55}$$

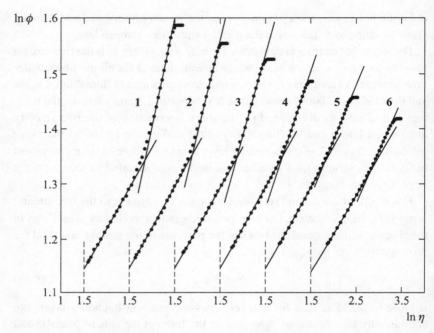

Figure 8.9. The mean velocity profiles in bilogarithmic coordinates in the series of
experiments of Marušić (1995) for $U = 30$ m/s and an adverse pressure gradient.
1: $Re = 19\,133$, $\ln Re_\Lambda = 8.83$, $P = 7.04 \times 10^{-3}$, $\beta = 0.388$. 2: $Re = 16\,584$,
$\ln Re_\Lambda = 10.18$, $P = 5.79 \times 10^{-3}$, $\beta = 0.346$. 3: $Re = 14\,208$, $\ln Re_\Lambda = 10.20$,
$P = 4.2 \times 10^{-3}$, $\beta = 0.306$. 4: $Re = 10\,997$, $\ln Re_\Lambda = 10.31$, $P = 2.86 \times 10^{-3}$,
$\beta = 0.247$. 5: $Re = 8588$, $\ln Re_\Lambda = 10.323$, $P = 1.75 \times 10^{-3}$, $\beta = 0.207$. 6:
$Re = 6430$, $\ln Re_\Lambda = 10.51$, $P = 0$, $\beta = 0.190$. The 'chevron' structure of the
profiles is clearly seen and regions I and II are clearly distinguishable.

We come to the conclusion that the power β and the coefficient B depend upon
two dimensionless governing parameters: Re and P.

So, we have arrived at a new model of the turbulent boundary layer at large
Reynolds numbers. According to the basic assumption of the models proposed
earlier by Clauser (1956) and by Coles (1956), which are widely accepted and
used, the transition from the flow in the wall region, described by the von
Kármán–Prandtl logarithmic law, to the external flow is smooth. According to
our model presented above, if the turbulence in the external flow is low then the
intermediate region between the viscous sublayer and the external flow consists
of two self-similar structures separated by a sharp boundary.

Thus, it may be said that, when processed, the experimental data published
over the last 30 years (see the references mentioned above) support the new
model.

It seems appropriate to finish this chapter with the words of Andrey Nikolae-vich Kolmogorov. At the end of his life, surveying the beginning of his work in turbulence, he said (see Kolmogorov 1991):

> It became clear for me that it is unrealistic to have a hope for the creation of a pure theory [of the turbulent flows of fluids and gases] closed in itself. Due to the absence of such a theory we have to rely upon the hypotheses obtained by processing of the experimental data. . . . I did not perform experimental work myself, but I spent a lot of energy on calculations and graphical processing of the data of other researchers.

(Note a parallel with von Kármán's introduction to his lecture, see p. 138.)

Twenty years have passed since these words were said, and 65 years since Kolmogorov started his work in turbulence. Unfortunately, very little has changed during these years as far as the possibility of constructing a pure theory closed in itself is concerned. However, more experimental data has appeared, and it is possible now, if needed, to modify the previous hypotheses and special theories based on them.

References

Aronson, D.G. (1986). *The Porous Medium Equation. Some Problems of Non-linear Diffusion*. Lecture Notes in Mathematics, 1224, Springer-Verlag, New York.

Barenblatt, G.I. (1952). On some unsteady motions of fluids and gases in a porous medium. *Prikl. Mat. Mekh.* **16** (1), 67–78.

Barenblatt, G.I. (1959a). On the equilibrium cracks formed in brittle fracture. *Appl. Math. Mech. (PMM)* **23** (3), 434–44; (4), 706–21; (5), 893–900.

Barenblatt, G.I. (1959b). The problem of thermal self-ignition. In: I.M. Gelfand, *Some Problems of the Theory of Quasi-Linear Equations*, Russian Mathematical Surveys, vol. 14 (2), pp. 137–42.

Barenblatt, G.I. (1962). Mathematical theory of equilibrium cracks in brittle fracture. *Adv. Appl. Mech.* **7**, 55–129.

Barenblatt, G.I. (1993). Scaling laws for fully developed turbulent shear flows. Part 1: Basic hypotheses and analysis. *J. Fluid Mech.* **248**, 513–20.

Barenblatt, G.I. (1994). *Scaling Phenomena in Fluid Mechanics*. Cambridge University Press.

Barenblatt, G.I. (1996). *Scaling, Self-similarity, and Intermediate Asymptotics*. Cambridge University Press.

Barenblatt, G.I. (1999). Scaling laws for turbulent wall-bounded shear flows at very large Reynolds numbers. *J. Eng. Math.* **36**, 361–84.

Barenblatt, G.I. and Chorin, A.J. (1996). Small viscosity asymptotics for the inertial range of local structure and for the wall region of wall-bounded turbulent shear flows. *Proc. US Nat. Acad. Sci.* **93**, 6749–52.

Barenblatt, G.I. and Chorin, A.J. (1997). Scaling laws and vanishing viscosity limits for wall-bounded shear flows and for local structure in developed turbulence. *Comm. Pure Appl. Math.* **50**, 381–98.

Barenblatt, G.I. and Monin, A.S. (1983). Similarity principles for the biology of pelagic animals. *Proc. Nat. Acad. Sci. USA* **80** (6), 3540–2.

Barenblatt, G.I. and Prostokishin, V.M. (1993). Scaling laws for fully developed turbulent shear flows. Part 2: Processing of experimental data. *J. Fluid Mech.* **248**, 521–9.

Barenblatt, G.I. and Vishik, M.I. (1956). On the finite speed of propagation in the problems of unsteady filtration of fluid and gas in a porous medium. *Appl. Math. Mech. (PMM)* **20** (4), 411–17.

Barenblatt, G.I. and Zeldovich, Ya.B. (1957). On the dipole-type solution in the problem of a polytropic gas flow in a porous medium. *Appl. Math. Mech. (PMM)* **21** (5), 718–20.

Barenblatt, G.I. and Zeldovich, Ya.B. (1971). Intermediate asymptotics in mathematical physics. *Russian Mathematical Surveys* **26** (2), 45–61.

Barenblatt, G.I. and Zeldovich, Ya.B. (1972). Self-similar solutions as intermediate asymptotics. *Ann. Rev. Fluid Mech.* **4**, 285–312.

Barenblatt, G.I., Chorin, A.J. and Prostokishin, V.M. (1997). Scaling laws in fully developed turbulent pipe flow. *Appl. Mech. Rev.* **50**, 413–29.

Barenblatt, G.I., Chorin, A.J. and Prostokishin, V.M. (2000a). Self-similar intermediate structures in turbulent boundary layers at large Reynolds numbers. *J. Fluid Mech.* **410**, 263–83.

Barenblatt, G.I., Chorin, A.J. and Prostokishin, V.M. (2000b). Characteristic length scale of the intermediate structure in zero-pressure-gradient boundary layer flow. *Proc. US Nat. Acad. Sci.* **97**, 3799–802.

Barenblatt, G.I., Chorin, A.J. and Prostokishin, V.M. (2002). A model of turbulent boundary layer with a non-zero pressure gradient. *Proc. US Nat. Acad. Sci.* **99**, 5772–6.

Barenblatt, G.I., Entov, V.M. and Ryzhik, V.M. (1990). *Theory of Fluid Flows Through Natural Rocks*. Kluwer Academic Publishers, Dordrecht, Boston, London.

Barenblatt, G.I., Chorin, A.J., Hald, O.H. and Prostokishin, V.M. (1997). Structure of the zero-pressure-gradient turbulent boundary layer. *Proc. US Nat. Acad. Sci.* **94**, 7817–19.

Batchelor, G.K. (1967). *An Introduction to Fluid Dynamics*. Cambridge University Press.

Batchelor, G.K. (1947). Kolmogoroff's theory of locally isotropic turbulence. *Proc. Cambridge Phil. Soc.* **43** (4), 533–59.

Batchelor, G.K. (1996). *The Life and Legacy of G.I. Taylor*. Cambridge University Press, Cambridge.

Bear, J. (1972). *Dynamics of Fluids in Porous Media*. American Elsevier, New York.

Benbow, J.J. (1960). Cone cracks in fused silica. *Proc. Phys. Soc.* **B75**, 697–9.

Bertsch, M., Dal Passo, R. and Ughi, M. (1986). Nonuniqueness and irregularity results for a nonlinear degenerate parabolic equation. In: *Proceedings of the Microprogram*, Berkeley, pp. 147–59.

Bertsch, M., Dal Passo, R. and Ughi, M. (1992). Nonuniqueness of solutions of a degenerate parabolic equation. *Annali di Matematica Pura ed Applicata (IV)* **CLXI**, 57–81.

Birkhof, G. (1960). *Hydrodynamics, A Study in Logic, Fact, and Similitude*, second edition. Princeton University Press.

Blasius, H. (1908). Grenzschichten in Flüssigkeit mit kleiner Reibung. *Z. Math. und Phys.* **56**, 1–37.

Bluman, G.W. and Cole, J.D. (1974). *Similarity Methods for Differential Equations*. Springer-Verlag, New York etc.

Bogolyubov, N.N. and Shirkov, D.V. (1955). On the renormalization group in quantum electrodynamics. *Doklady USSR Ac. Sci.* **103** (2), 203–6.

Bogolyubov, N.N. and Shirkov, D.V. (1959). *Introduction to the Theory of Quantized Fields*. Wiley Interscience, New York, London.

Bose, E. and Bose, M. (1911). Über die Turbulenzreibung verschiedener Flüssigkeiten. *Phys. Zeit.* **12** (4), 126–35.

Bose, E. and Rauert, D. (1909). Experimentalbeitrag zur Kenntnis der turbulenten Flüssigkeitsreibung. *Phys. Zeit.* **10** (12), 406–9.

Bridgman, P.W. (1931). *Dimensional Analysis.* Yale University Press, New Haven.

Broberg, K.B. (1999). *Cracks and Fracture.* Academic Press, San Diego etc.

Brown, G. (1991). ARPA–URI Proposal, Princeton University.

Chen, L.-Y. and Goldenfeld, N. (1992). Renormalization-group theory for the propagation of a turbulent burst. *Phys. Rev.* **A45** (8), 5572–4.

Chen, L.-Y., Goldenfeld, N. and Oono, Y. (1991). Renormalization-group theory for the modified porous-medium equation. *Phys. Rev.* **A44** (10), 6544–50.

Chernyi, G.G. (1961). *Introduction to Hypersonic Flows* (transl. R.F. Probstein). Academic Press, New York.

Chorin, A.J. (1988). Scaling laws in the vortex lattice model of turbulence. *Comm. Math. Phys.* **114**, 167–76.

Chorin, A.J. (1994). *Vorticity and Turbulence.* Springer-Verlag, New York.

Chorin, A.J. (1998). New perspectives in turbulence. *Quart. J. Appl. Math.* **XIV** (4), 767–85.

Clark, R.W. (1961). *The Birth of the Bomb. The Untold Story of Britain's Part in the Weapon That Changed the World.* Phoenix House, London.

Clauser, F.H. (1956). The turbulent boundary layer. *Adv. Appl. Mech.* **4**, Academic Press, New York, pp. 2–52.

Cole, J.D. (1968). *Perturbation Methods in Applied Mathematics.* Blaisdell, Toronto, London.

Coles, D.E. (1956). The law of the wake in a turbulent boundary layer. *J. Fluid Mech.* **1** (3), 191–226.

Daniell, P.J. (1930). The theory of flame motion. *Proc. Roy. Soc.* **A126**, 393–402.

Erm, L.P. and Joubert, P.N. (1991). Low Reynolds-number turbulent boundary layers. *J. Fluid Mech.* **230**, 1–44.

Entov, V.M. (1994). Private communication.

Fernholz, H.H. and Finley, P.J. (1996). The incompressible zero-pressure gradient turbulent boundary layer: an assessment of the data. *Progr. Aeros. Sci.* **32**, 245–311.

Fisher, R.A. (1937). The wave of advance of advantageous genes. *Ann. Eugenics* **7**, 355–69.

Friedrichs, K.O. (1966). *Special Topics in Fluid Dynamics.* Gordon and Breach, New York, London, Paris.

Gardner, C.S.J., Greene, J.M., Kruskal, M.D. and Miura, R.M. (1967). A method for solving the Korteweg–de Vries equation. *Phys. Rev. Lett.* **19**, 1095–7.

Gell-Mann, M. and Low, F.E. (1954). Quantum electrodynamics at small distances. *Phys. Rev* **95**, 1300–12.

Germain, P. (1986). *Mécanique,* Ecole Polytechnique, Ellipses, Paris.

Goldenfeld, N. (1989). The approach to equilibrium: scaling and the renormalization group. Lecture by invitation at the Conference on Nonlinear Phenomena, Moscow, USSR Ac. Sci., 19–22 September.

Goldenfeld, N. (1992). *Lectures on Phase Transitions and the Renormalization Group.* Perseus Books, Reading, Mass.

Goldenfeld, N. and Oono, Y. (1991). Renormalization group theory for two problems in linear continuum mechanics. *Physica A* **177**, 213–19.

Goldenfeld, N., Martin, O. and Oono, Y. (1989). Intermediate asymptotics and renormalization group theory. *J. Scient. Comput.* **4**, 355–72.

Goldenfeld, N., Martin, O. and Oono, Y. (1991). Asymptotics of partial differential equations and the renormalization group. In: S. Tanvera (ed.), *Proceedings NATO Advanced Research Workshop on Asymptotics Beyond all Orders*, La Jolla, Plenum Press.

Goldenfeld, N., Martin, O., Oono, Y. and Liu, F. (1990). Anomalous dimensions and the renormalization group in a nonlinear diffusion process. *Phys. Rev. Lett.* **65** (12), 1361–4.

Golitsyn, G.S. (1973). *Introduction to the Dynamics of Planetary Atmospheres.* Gidrometeoizdat, Leningrad.

Guderley, K.G. (1942). Starke kügelige und zylindrische Verdichtuagsstösse in der Nähe des Kügelurittelpuukte bzw. der Zylinderachse. *Luftfahrtforschung* **19** (9), 302–12.

Hancock, P.E. and Bradshaw, P. (1989). Turbulence structure of a boundary layer beneath a turbulent free stream. *J. Fluid Mech.* **205**, 45–76.

Hinch, E.J. (1991). *Perturbation Methods.* Cambridge University Press.

Izakson, A. (1937). Formula for the velocity distribution near a wall. *J. Exp. Theor. Physics* **7**, 919–24.

Kadanoff, L.P. (1966). Scaling laws for Ising model near T_c. *Physics* **2** (6), 263–72.

Kadanoff, L.P. *et al.* (1967). Static phenomena near critical points: theory and experiment. *Rev. Mod. Phys.* **39** (2), 395–431.

Kalashnikov, A.S. (1987). Some problems of qualitative theory of non-linear second-order parabolic equations. *Russian Mathematical Surveys* **42**, 169–222.

Kapitza, S.P. (1966). A natural system of units in classical electrodynamics and electronics. *Sov. Phys. Uspekhi* **9**, 184.

Kolmogorov, A.N. (1941). The local structure of turbulence in incompressible fluids at very high Reynolds numbers. *Doklady, USSR Ac. Sci.* **30** (4), 299–303.

Kolmogorov, A.N. (1991). *Selected works*, vol. 1, *Mathematics and Mechanics*. V.M. Tikhomirov (ed). Kluwer, Dordrecht etc.

Kolmogorov, A.N., Petrovsky, I.G. and Piskunov, N.S. (1937). Investigation of the diffusion equation connected with an increasing amount of matter and its application to a biological problem. *Bull. MGU* **A1** (6), 1–26.

Krogstad, P.-Å. and Antonia, R.A. (1999). Surface roughness effects in turbulent boundary layers. *Exp. Fluids* **27**, 450–60.

Landau, L.D. and Lifshitz, E.M. (1987). *Fluid Mechanics*, second edition. Pergamon Press, London, New York.

Latter, R. (1955). Similarity solution for a spherical shock wave. *J. Appl. Phys.* **26**, 954–60.

Lax, P.D. (1968). Integrals of nonlinear equations of evolution and solitary waves. *Comm. Pure Appl. Math.* **21** (5), 467–90.

Lighthill, M.J. (1968). Turbulence. In: D.M. McDowell and J.D. Jackson (eds.), *Osborne Reynolds and Engineering Science Today*, Manchester University Press, Manchester, pp. 83–146.

Lockwood Taylor, J. (1955). An exact solution of the spherical blast wave problem. *Phil. Mag.* **46**, 317–20.

Mandelbrot, B. (1975). *Les objects fractals: forme, hasard et dimension.* Flammarion, Paris.

Mandelbrot, B. (1977). *Fractals, Form, Chance and Dimension.* W.H. Freeman and Co., San Francisco.

Mandelbrot, B. (1982). *The Fractal Geometry of Nature.* W.H. Freeman and Co., San Francisco.

Marušić, I. and Perry, A.E. (1995). A wall-wake model for the turbulence structure of boundary layers. Part 2. Further experimental support. *J. Fluid Mech.* **298**, 389–407. (http://www.mame.mu.oz.au/ivan)

McMahon, T.A. (1971). Rowing: a similarity analysis. *Science* **173**, 23 July 1971, 349–51.

Migdal, A.B. (1977). *Qualitative Methods in Quantum Theory.* W.A. Benjamin, Reading, Mass.

Millikan, C.B. (1939). A critical discussion of turbulent flows in channels and circular tubes. In: J.P. Den Hartog and H. Peters (eds), *Proc. 5th Int. Congress Appl. Mech.* Cambridge, Mass., pp. 386–92.

Monin, A.S. and Yaglom, A.M. (1971). *Statistical Fluid Mechanics. Mechanics of Turbulence*, vol. 1. MIT Press, Cambridge, Mass., London.

Murray, J.D. (1977). *Lectures on Nonlinear Differential Equation Models in Biology.* Clarendon Press, Oxford.

Nagib, H. and Hites, M. (1995). High Reynolds number boundary layer measurements in the NDF. AIAA paper 95–0786.

Naguib, A.N. (1992). Inner- and outer-layer effects on the dynamics of a turbulent boundary layer. Ph.D. thesis, Illinois Institute of Technology.

Nikolaeva, G.G. (1975). Metabolism intensity and size–weight characteristics of the *Rhithropanopeus harisii tridentatus* crab from the Caspian Sea. *Oceanology USSR* **15** (1), 140–2.

Nikuradze, J. (1932). Gesetzmässigkeiten der turbulenten Strömung in glatten Röhren. VDI Forschungscheftg no. 356.

Obukhov, A.M. (1941). On the distribution of energy in the spectrum of a turbulent flow. *Doklady, USSR Ac. Sci.* **32** (1), 22–4.

Oleynik, O.A. (1957). Discontinuous solutions of nonlinear differential equations. *Uspekhi Mat. Nauk.* **12** (3) (75th issue), 3–73.

Oleynik, O.A., Kalashnikov, A.S. and Chzhou Yui-lin (1958). The Cauchy problem and boundary problems for equations of the type of unsteady filtration. *Izvest. USSR Ac. Sci. Ser. Mat.* **22**, 667–704.

Olver, P.J. (1993). *Applications of Lie Groups to Differential Equations.* Springer-Verlag, New York etc.

Österlund, J.M., Johansson, A.V., Nagib, H.M. and Hites, M.H. (1999). Wall shear stress measurements in high Reynolds number boundary layers from two facilities. AIAA paper 99–3814.

Panton, R.L. (2002), Evaluation of the Barenblatt–Chorin–Prostokishin power law for turbulent boundary layers. *Physics of Fluids* **14** (5), 1806–8.

Patashinsky, A.Z. and Pokrovsky, V.L. (1966). On the behaviour of ordering systems near the phase transition point. *J. Exp. Theor. Phys.* **50** (2), 439–47.

Pattle, R.E. (1959). Diffusion from an instantaneous point source within a concentration-dependent coefficient. *Quart. J. Mech. Appl. Math.* **12**, 407–9.

Perry, A.E., Hafer, S. and Chong, M.S. (2001). A possible reinterpretation of the Princeton superpipe data. *J. Fluid Mech.* **439**, 395–401.

Polubarinova-Kochina, P.Ya. (1962). *Theory of Groundwater Movement.* Princeton University Press.

Prandtl, L. (1904). Über Flüssigkeitsbewegung bei sehr kleinen Reibung. Verhandlungen III. *Intern. Math. Kongress, Leipzig*, 484–91.

Prandtl, L. (1932). Zur turbulenten Strömung in Röhren und längs Platten. *Ergebn. Aerodyn. Versuchsanstalt, Göttingen* **B4**, 18–29.

Richardson, L.F. (1961). The problem of contiguity: an appendix of statistics of deadly quarrels. *General Systems Year Book* **6**, 139–87.

Roesler, F. (1956). Brittle fracture near equilibrium. *Proc. Phys. Soc.* **B69**, 981–92.

Schlichting, H. (1968). *Boundary Layer Theory*, sixth edition. McGraw-Hill, New York.

Sedov, L.I. (1946). Propagation of strong shock waves. *Prikl. Mat. Mekh.* **10**, 241–50 (Pergamon Translations no. 1223).

Sedov, L.I. (1959). *Similarity and Dimensional Methods in Mechanics.* Academic Press, New York.

Shushkina, E.A., Kus'micheva, V.I. and Ostapenko, L.A. (1971). Energy equivalent of body mass, respiration, and calorific value of mysids from the Sea of Japan. *Oceanology USSR* **11** (6), 880–3.

Spurk, J.H. (1997). *Fluid Mechanics.* Springer-Verlag, New York.

Staniukovich, K.P. (1960). *Unsteady Motion of Continuous Media.* Pergamon Press, New York.

Stückelberg, E.C.G. and Peterman, A. (1953). La normalisation des constantes dans la théorie des quanta. *Helvetica Physica Acta* xxvi, 499–520.

Taffanel, M. (1913). Sur la combustion des mélanges gazeux et les vitesses de réaction. *C.R. Ac. Sci. Paris* **157**, 714–17.

Taffanel, M. (1914). Sur la combustion des mélanges gazeux et les vitesses de réaction. *C.R. Ac. Sci. Paris* **158**, 42–5.

Taylor, G.I. (1941). The formation of a blast wave by a very intense explosion. Report RC–210, 27 June 1941, Civil Defence Research Committee.

Taylor, G.I. (1950a). The formation of a blast wave by a very intense explosion. I: Theoretical discussion. *Proc. Roy. Soc.* **A201**, 159–74.

Taylor, G.I. (1950b). The formation of a blast wave by a very intense explosion. II: The atomic explosion of 1945. *Proc. Roy. Soc.* **A201**, 175–86.

Taylor, G.I. (1963). *Scientific Papers*, G.K. Batchelor (ed.), vol. 3, *Aerodynamics and the Mechanics of Projectiles and Explosions.* Cambridge University Press.

Töpfer, C. (1912). Ämerkungen zu dem Aufsatz von H. Blasius "Grenzschichten in Flüssigkeit mit kleiner Reibung". *Z. Math. Phys.* **60**, 397.

Van Dyke, M. (1975). *Perturbation Methods in Fluid Mechanics*, second edition. Parabolic Press, Stanford.

Van Dyke, M. (1982). *An Album of Fluid Motion.* Parabolic Press, Stanford.

von Kármán, Th. (1930). Mechanische Ähnlichkeit und Turbulenz. In: C.W. Oseen and W. Weibull (eds.), *Proc. 3rd Int. Congr. Appl. Mech.*, Stockholm, AB Sveriges Lifografska Tryckenier, vol. 1, pp. 85–93.

von Kármán, Th. (1957). *Aerodynamics.* Cornell University Press, Ithaca.

von Koch, H. (1904). Sur une courbe continue sans tangente obtenue par une construction géometrique élémentaire. *Arkiv Mat. Astron. Fys.* **2**, 681–704.

von Mises, R. (1941). Some remarks on the laws of turbulent motion in tubes. In: *Th. von Kármán, Anniversary Volume.* CalTech Press, Pasadena, pp. 317–27.

von Neumann, J. (1941). The point source solution. National Defence Research Committee, Div. B, Report AM–9, 30 June 1941.

von Neumann, J. (1963). The point source solution. In: A.H. Taub (ed.), *Collected Works*, Vol. VI, 219–37. Pergamon Press, Oxford, New York, London, Paris.

von Weizsäcker, C.F. (1954). Genäherte Darstellung starker instationärer Storswellen durch Homologie-Lösungen. *Z. Naturforschung* **9A**, 269–75.

Wilson, K. (1971). Renormalization group and critical phenomena, I, II. *Phys. Rev.* **B4** (9), 3174–83, 3184–205.

Zagarola, M.V. (1996). Mean flow scaling in turbulent pipe flow. Ph.D. thesis, Princeton University.

Zagarola, M.V., Smits, A.J., Orszag, S.A. and Yakhot, V. (1996). Experiments in high Reynolds number turbulent pipe flow. AIAA paper 95–0654, Reno, Nev.

Zeldovich, Ya.B. (1946). *Theory of Shock Waves and an Introduction to Gas Dynamics*, Publishing House of USSR Ac. Sci., Moscow.

Zeldovich, Ya.B. (1948). On the theory of flame propagation. *Zhurn. Fiz. Khimii* **22** (1), 27–48.

Zeldovich, Ya.B. (1956). The motion of a gas under the action of short term pressure shock. *Sov. Phys. Acoustics* **2**, 25–35.

Zeldovich, Ya.B. and Frank-Kamenetsky, D.A. (1938a). Theory of uniform propagation of flames. *Doklady USSR Ac. Sci.* **19** (2), 697–7.

Zeldovich, Ya.B. and Frank-Kamenetsky, D.A. (1938b). Theory of uniform propagation of flames. *Zhurn. Fiz. Khimii* **12** (1), 100–5.

Zeldovich, Ya.B. and Kompaneets, A.S. (1950). On the theory of propagation of heat with thermal conductivity depending on temperature. In *Collection of Papers Dedicated to the 70th Birthday of A.F. Ioffe*, pp. 61–71. Izd. Akad. Nauk. USSR, Moscow.

Zeldovich, Ya.B. and Raizer, Yu.P. (1966). *Physics of Shock Waves and High Temperature Hydrodynamic Phenomena*, vol. I. Academic Press, New York, London.

Zeldovich, Ya.B. and Raizer, Yu.P. (1967). *Physics of Shock Waves and High Temperature Hydrodynamic Phenomena*, vol. II. Academic Press, New York, London.

Zeldovich, Ya.B., Barenblatt, G.I., Librovich, V.B. and Makhviladze, G.M. (1985). *The Mathematical Theory of Combustion and Explosions.* Consultants Bureau, New York, London.

Index

Printed in the United States
By Bookmasters